Connectomics

Connectomics
Applications to Neuroimaging

Edited by

Brent C. Munsell

Guorong Wu

Leonardo Bonilha

Paul J. Laurienti

Academic Press is an imprint of Elsevier
125 London Wall, London EC2Y 5AS, United Kingdom
525 B Street, Suite 1650, San Diego, CA 92101, United States
50 Hampshire Street, 5th Floor, Cambridge, MA 02139, United States
The Boulevard, Langford Lane, Kidlington, Oxford OX5 1GB, United Kingdom

Notices
Knowledge and best practice in this field are constantly changing. As new research and experience
broaden our understanding, changes in research methods, professional practices, or medical treatment
may become necessary.

Practitioners and researchers must always rely on their own experience and knowledge in evaluating and
using any information, methods, compounds, or experiments described herein. In using such information
or methods they should be mindful of their own safety and the safety of others, including parties for
whom they have a professional responsibility.

To the fullest extent of the law, neither the Publisher nor the authors, contributors, or editors, assume any
liability for any injury and/or damage to persons or property as a matter of products liability, negligence
or otherwise, or from any use or operation of any methods, products, instructions, or ideas contained in
the material herein.

Library of Congress Cataloging-in-Publication Data
A catalog record for this book is available from the Library of Congress

British Library Cataloguing-in-Publication Data
A catalogue record for this book is available from the British Library

ISBN 978-0-12-813838-0

For information on all Academic Press publications
visit our website at https://www.elsevier.com/books-and-journals

Publisher: Mara Conner
Acquisition Editor: Tim Pitts
Editorial Project Manager: Leticia Lima
Production Project Manager: R.Vijay Bharath
Cover Designer: Mark Rogers

Typeset by SPi Global, India

Contents

CHAPTER 9 **Characterizing Dynamic Functional Connectivity Using Data-Driven Approaches and its Application in the Diagnosis of Alzheimer's Disease** **181**

Yingying Zhu, Xiaofeng Zhu, Minjeong Kim, Daniel Kaufer, Paul J. Laurienti, Guorong Wu

CHAPTER 10 **Toward a more Integrative Cognitive Neuroscience of Episodic Memory** **199**

Matthew L. Stanley, Benjamin R. Geib, Simon W. Davis

Contributors

Vince D. Calhoun
The Mind Research Network; Department of Electrical and Computer Engineering, University of New Mexico, Albuquerque, NM, United States

Jessica R. Cohen
Department of Psychology and Neuroscience, University of North Carolina at Chapel Hill, Chapel Hill, NC, United States

R. Todd Constable
Interdepartmental Neuroscience Program; Department of Radiology and Biomedical Imaging; Department of Neurosurgery, Yale University School of Medicine, New Haven, CT, United States

Dale Dagenbach
Department of Psychology, Wake Forest University, Winston-Salem, NC, United States

Simon W. Davis
Department of Neurology, Center for Cognitive Neuroscience, Duke University, Durham, NC, United States

Emily L. Dennis
Imaging Genetics Center, Mark and Mary Stevens Neuroimaging and Informatics Institute, Keck School of Medicine of the University of Southern California, Marina del Rey, CA, United States

Xiaoyu Ding
Neuroimaging Research Branch, National Institute on Drug Abuse, Intramural Research Program, National Institutes of Health, Baltimore, MD, United States

Benjamin R. Geib
Department of Psychology and Neuroscience, Center for Cognitive Neuroscience, Duke University, Durham, NC, United States

Teague R. Henry
Department of Psychology and Neuroscience, University of North Carolina at Chapel Hill, Chapel Hill, NC, United States

Corey Horien
Interdepartmental Neuroscience Program, Yale University School of Medicine, New Haven, CT, United States

Neda Jahanshad
Imaging Genetics Center, Mark and Mary Stevens Neuroimaging and Informatics Institute, Keck School of Medicine of the University of Southern California, Marina del Rey, CA, United States

Daniel Kaufer
Department of Neurology, University of North Carolina at Chapel Hill, Chapel Hill, NC, United States

Minjeong Kim
Department of Radiology and BRIC, University of North Carolina at Chapel Hill, Chapel Hill, NC, United States

Paul J. Laurienti
Department of Radiology, Wake Forest School of Medicine, Winston-Salem, NC, United States

Eva Mennigen
Department of Psychiatry and Biobehavioral Sciences, Semel Institute for Neuroscience and Human Behavior, University of California, Los Angeles, CA, United States

Luiz Pessoa
Department of Psychology and Maryland Neuroimaging Center, University of Maryland, College Park, MD, United States

Barnaly Rashid
Department of Psychiatry, Harvard Medical School, Boston, MA, United States

Thomas J. Ross
Neuroimaging Research Branch, National Institute on Drug Abuse, Intramural Research Program, National Institutes of Health, Baltimore, MD, United States

Dustin Scheinost
Department of Radiology and Biomedical Imaging; The Child Study Center, Yale University School of Medicine, New Haven, CT, United States

Matthew L. Stanley
Department of Psychology and Neuroscience, Center for Cognitive Neuroscience, Duke University, Durham, NC, United States

Vaughn R. Steele
Neuroimaging Research Branch, National Institute on Drug Abuse, Intramural Research Program, National Institutes of Health, Baltimore, MD, United States

Paul M. Thompson
Imaging Genetics Center, Mark and Mary Stevens Neuroimaging and Informatics Institute, Keck School of Medicine of the University of Southern California, Marina del Rey, CA, United States

Archana Venkataraman
Department of Electrical and Computer Engineering, Johns Hopkins University, Baltimore, MD, United States

Guorong Wu
Department of Radiology and BRIC, University of North Carolina at Chapel Hill, Chapel Hill, NC, United States

Yingying Zhu
School of Electrical and Computer Engineer, Cornell University, Ithaca, NY, United States

Xiaofeng Zhu
Department of Radiology, Perelman School of Medicine, University of Pennsylvania, Philadelphia, PA, United States

Autism spectrum disorders: Unbiased functional connectomics provide new insights into a multifaceted neurodevelopmental disorder

1

Archana Venkataraman

Department of Electrical and Computer Engineering, Johns Hopkins University,
Baltimore, MD, United States

CHAPTER OUTLINE

INTRODUCTION

Autism spectrum disorder (ASD) affects an estimated 1 in 68 children in the United States, often with devastating effects on both patients and family members (Centers for for Disease Control, 2012; Leslie and Martin, 2007; Stuart and McGrew, 2009). From a neuroscientific perspective, ASD cannot be viewed as a single unified brain dysfunction (Aoki et al., 2013; Waterhouse and Gilberg, 2014); rather, it manifests through a series of distributed interactions across the brain (Cherkassky et al., 2006; Geschwind and Konopka, 2009; Sullivan et al., 2014). Behaviorally, ASD is characterized by blunted sociocommunicative skill and awareness across multiple sensory domains (Kanner, 1943; Pelphrey et al., 2014), coupled with stereotyped patterns

Connectomics. https://doi.org/10.1016/B978-0-12-813838-0.00001-7

of behaviors (Americal Psychiatric Association, 2013). However, the manifestation and severity of these clinical symptoms vary considerably across individuals and over the lifespan of each patient. In short, ASD is a complex and multifaceted disorder, and despite ongoing efforts, we have a limited understanding of its origin and pathogenesis. (Gabrieli-Whitfield et al., 2009; Hernandez et al., 2015; Sullivan et al., 2014).

Among its diverse behavioral presentations, social and language dysfunctions are considered hallmark and unifying features of ASD (Baron-Cohen et al., 1999; Kanner, 1943; Pelphrey et al., 2014). Social impairments are apparent in both verbal and nonverbal domains, and they manifest across simple (e.g., shared gaze) and complex (e.g., back-and-forth conversation) behaviors. On the language side, patients with ASD have notable difficulties with the production and interpretation of human speech (Globerson et al., 2015; Grossman et al., 2010). Because these deficits emerge within the first years of life, one popular theory in the field suggests that ASD alters both the structural and functional development of the brain via experience-dependent processes (Courchesne and Pierce, 2005; Geschwind and Levitt, 2007; Just et al., 2012; Melillo and Leisman, 2011). Functional magnetic resonance imaging (fMRI) is one of the most popular tools for studying neurological changes in ASD. For example, an investigation of speech processing in sleeping 2- to 3-year-old children found reduced activity in brain regions associated with language comprehension (Redcay and Courchesne, 2008). Likewise, a study of verbal fluency found atypical hemispheric lateralization in ASD (Kleinhans et al., 2008). Other fMRI studies have revealed significant changes in neural activity related to reward processing (Scott-Van Zeeland et al., 2010; Schmitz et al., 2008), joint attention (Belmonte and Yurgelun-Todd, 2003; Williams et al., 2005), and working memory (Koshino et al., 2005, 2008). Although valuable, it is worth emphasizing that these paradigms are designed to trigger specific neural activation using a narrow range of experimental stimuli. For this reason, one can argue that task fMRI studies do not capture naturalistic and whole-brain interactions.

This book chapter highlights the promise of functional connectomics in the study of ASD. Unlike task-based paradigms, functional connectomics allows us to quantify the synchrony between brain regions at both local and long-range scales. This flexibility offers a holistic perspective of ASD across multiple brain systems. Likewise, the absence of external stimuli allows us to focus on intrinsic or steady-state communication patterns. "Functional Connectomics as a Window into ASD" section of this chapter summarizes prior work in the field, from simple seed-based analyses to more complex network models. "An Unbiased Bayesian Framework for Functional Connectomics" section introduces our novel Bayesian framework to extract the altered subnetworks associated with ASD. In "Multisite Network Analysis of Autism" section, we evaluate this model on a multisite study of autism. "Toward Characterizing Patient Heterogeneity" section presents a new extension of our framework that incorporates patient heterogeneity. Finally, we conclude with some general recommendations for future work in the field.

FUNCTIONAL CONNECTOMICS AS A WINDOW INTO ASD

Functional connectomics provides a unique glimpse into the steady-state organization of the brain. It is based on the underlying assumption that two regions, which reliably coactivate, are more likely to participate in similar neural processes than two uncorrelated or anticorrelated regions (Buckner and Vincent, 2007; Fox and Raichle, 2007; Van Dijk et al., 2010; Venkataraman et al., 2009). Over the past decade, functional connectomics has become ubiquitous in the study of neurological disorders, such as schizophrenia, epilepsy, and autism (DiMartino et al., 2014; Liang et al., 2006; Stufflebeam et al., 2011). From a practical standpoint, these functional relationships are often evaluated in resting-state fMRI (rsfMRI), which does not require patients to complete potentially challenging experimental paradigms. From a neuroscientific standpoint, group-level changes in the functional architecture of the brain may shed light on the etiological mechanisms of a disorder.

Univariate tests have historically been the standard approach to isolate the altered functional connectivity patterns in ASD (Cherkassky et al., 2006; Kennedy and Courchesne, 2008b). These methods identify statistical differences in pairwise similarity measures, such as Pearson correlation coefficients or seed-based correlation maps, as representative biomarkers of ASD. Perhaps the most notable findings have been a consistent reduction in interregional connections, particularly between the frontal and posterior lobes (Hull et al., 2016; Just et al., 2004, 2012), and connectivity differences linked to the default mode network (DMN), which activates during self-reflective processes (Buckner et al., 2008; Kennedy and Courchesne, 2008a; Padmanabhana et al., 2017). Interestingly, many studies have reported greater intraregional connectivity in some ASD subpopulations (Delmonte et al., 2013; DiMartino et al., 2014), which may be linked to enhanced sensory perception. Unfortunately, univariate results are wildly inconsistent across the ASD literature. One contributing factor to their low test-retest reliability is that, by construction, univariate tests ignore crucial dependencies across the brain (Venkataraman et al., 2010).

Graph models assume a structured relationship between the pairwise connectivity values to estimate surrogates of both functional specialization and functional integration (Achard and Bullmore, 2007; Bassett and Bullmore, 2009; Bullmore and Sporns, 2009; Rubinov and Sporns, 2010). For example, modularity and clustering coefficient quantify the interconnectedness of local processing units (functional specialization) (Meunier et al., 2009; Rubinov and Sporns, 2010; Sporns and Betzel, 2016), whereas average path length, global efficiency, and betweenness centrality quantify the reachability of each node in the network (functional integration) (Achard and Bullmore, 2007; Estrada and Hatano, 2008). Finally, the small-world architecture balances these competing influences (Tononi et al., 1994). The past 5 years has witnessed a proliferation in graph-theoretic studies of ASD. One interesting finding is a decrease in clustering coefficient and "hubness" across the brain, which suggests that, on average, ASD patients have a more random network organization than neurotypical controls (Itahashi et al., 2014). There has also been conflicting evidence to support the popular theory of local overconnectivity and long-range underconnectivity in ASD

(Takashi Itahashi et al., 2015; Keown et al., 2013; Redcay et al., 2013; Rudie and Dapretto, 2017). Although graph measures have provided some insight into ASD, they are markedly removed from the original network. Therefore, it is unclear what neural mechanisms contribute to these measures, whether group differences reflect a verifiable change in the underlying functional organization, or whether they stem from a confounding influence (Smith, 2012).

An alternate network approach is to decompose the rsfMRI time series into a collection of hidden sources in the brain. The increasingly popular independent component analysis (ICA) relies on statistical independence and non-Gaussianity to guide the network decomposition (Bell and Sejnowski, 1995; Calhoun et al., 2003; McKeown et al., 1998). When applied to rsfMRI data, ICA returns both a spatial map and a representative time series for each component/source (Calhoun et al., 2003, 2009). The anatomical organization of these components can be used to delineate different functional networks in the brain, and temporal fluctuations in the time series quantify the synchrony across networks. To a large extent, ICA studies for ASD focus on the similarity between selected ICA components (i.e., networks), such as the DMN (Assaf et al., 2010; Starck et al., 2013; Supekar et al., 2010), subcortical areas (Cerliani et al., 2015), the sensorimotor network (Nebel et al., 2014, 2016), and the prefrontal cortex (Starck et al., 2013). However, the main drawback of ICA is that it does not naturally generalize to multisubject or population level analyses (Calhoun et al., 2009).

Despite the breadth of analysis techniques, the previously discussed methods follow a similar two-step procedure for studying ASD: they first fit a connection- or graph-based model to the rsfMRI data and then identify group differences post hoc. In practice, this strategy tends to implicate distributed, and potentially unrelated, changes in functional connectivity across the brain. These isolated effects are difficult to interpret and are often missing crucial details about the functional architecture of the brain. To this end, we have developed a novel probabilistic framework that identifies *network-based differences* in functional connectivity. Our unique methodology extracts robust and clinically meaningful biomarkers of ASD from multisite connectivity data (Venkataraman et al., 2015). We also discuss a recently proposed extension of our model that incorporates a patient-specific measure of ASD severity into the Bayesian framework (Venkataraman et al., 2017).

AN UNBIASED BAYESIAN FRAMEWORK FOR FUNCTIONAL CONNECTOMICS

Given the growing perception of ASD as a system-level dysfunction (Courchesne and Pierce, 2005; Geschwind and Levitt, 2007), we hypothesize that the functional differences attributed to ASD reflect a set of *coordinated disruptions* in the brain. Although we do not specify a priori whether these disruptions occur within the same cognitive domain or whether they span multiple cognitive processes, we assume that the affected brain regions will communicate differently with other parts of the brain than if the disorder were not present. In the functional connectomics realm, this underlying assumption can be modeled by region hubs, which exhibit a large number of altered functional connections, compared with the neurotypical cohort. In the following, we refer to these region hubs as *disease foci* and the altered functional connectivity patterns as *canonical networks*.

Fig. 1 outlines the generative process. The connectivity differences in ASD are explained by a set of K nonoverlapping networks, where K is a user-specified parameter that controls the model complexity. We use a probabilistic framework to represent the interaction between regions that describe the effects of ASD. Here, *latent* variables specify a template organization of the brain, which we cannot directly access. Instead, we observe noisy measurements of the hidden structure via rsfMRI correlations.

As seen, our framework is based on hierarchical variable interactions. The multinomial variable R_i indicates whether region i is healthy ($R_i = 0$) or whether it is a disease focus in network k ($R_i = k$). The latent functional connectivity F_{ij} describes the group-wise coactivation between region i and region j in the neurotypical controls based on one of three states: positive synchrony ($F_{ij} = 1$), negative synchrony ($F_{ij} = -1$), and no coactivation ($F_{ij} = 0$). Notice that our discrete representation of latent functional connectivity is a notable departure from conventional analysis. Specifically, we assume that rsfMRI correlations fall into one of three general categories, and differences in bin assignment are the relevant markers of ASD. Our choice of three states is motivated by the rsfMRI literature. For example, most works specify a threshold to determine functionally connected areas, which corresponds to $F_{ij} = 1$ in our framework. On the other hand, although strong negative correlations do appear in rsfMRI data, there is no consensus about their origin and significance (Van Dijk et al., 2010). Therefore, we isolate negative connectivity (i.e., $F_{ij} = -1$) as a separate category. The latent functional connectivity $\bar{F}_{ij}$ of the ASD population is also tristate and is defined via four simple rules: (1) a connection between two disease foci in the same network k is always abnormal, (2) a connection between two foci in different networks is never abnormal, (3) a connection between two healthy regions is never abnormal, and (4) a connection between a healthy and a diseased region is abnormal with probability η. Ideally $\bar{F}_{ij} \neq F_{ij}$ for abnormal connections and $\bar{F}_{ij} = F_{ij}$ for healthy connections. However, due to noise, we assume that the latent templates can deviate from these rules with probability ε. Notice that condition 2 ensures that the K networks remain distinct, and conditions 3 and 4 impose an outward spreading topology on the altered pathways. Finally, the rsfMRI correlations B_{ij}^l for neurotypical subject l and $\bar{B}_{ij}^m$ for ASD patient m are sampled from Gaussian distributions whose mean and variance depend on the neurotypical and ASD functional templates, respectively. The beauty of our proposed hierarchical model is that we are able to isolate the effects of ASD within the latent structure, while simultaneously accounting for noise and subject variability via the data likelihood.

We derive a variational expectation-maximization (EM) algorithm (Jordan et al., 1999) to estimate both the latent posterior probability of each region label q_i and the nonrandom model parameters from the observed data. A full mathematical characterization of the model and optimization algorithm are given in our previous publications (Venkataraman et al., 2015, 2013a).

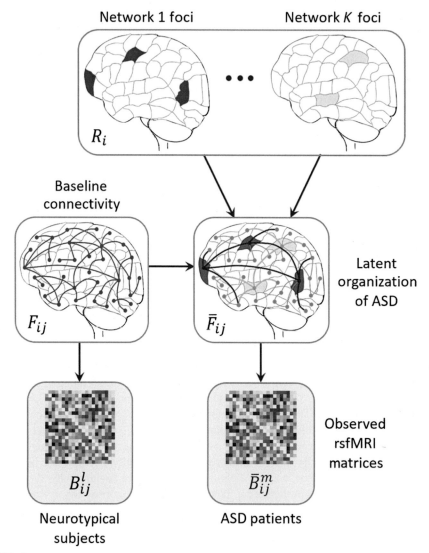

FIG. 1

Generative model of functional connectivity for ASD. Parcels correspond to regions in the brain, and lines denote pairwise functional connections. The label R_i indicates whether region i is healthy *(white)* or a focus in one of the K abnormal networks *(colored)*. These foci capture the most salient functional differences between patients and controls. The neurotypical template F_{ij} provides a baseline functional architecture of the brain, whereas the clinical template $\bar{F}_{ij}$ specifies the functional differences attributed to ASD. The subject rsfMRI correlations $\{B_{ij}^l\}$ and $\{\bar{B}_{ij}^m\}$ are noisy observations of the latent functional templates.

Our methodology circumvents the interpretability challenges of the simple statistical analyses that currently dominate the clinical neuroscience literature. For example, univariate tests are commonly used to identify group-wise differences in pairwise correlation values. However, the bulk of our knowledge about the brain is organized around regions and not the connections between them. Moreover, connection-based results are nearly impossible to verify through direct stimulation. On the flipside, popular graph measures, such as modularity and small-worldness (Bassett and Bullmore, 2006; Honey et al., 2009; Rubinov and Sporns, 2010) collapse the rich network structures onto scalar values. As a result, we cannot tie statistical differences to a concrete etiological mechanism. In contrast, we explicitly model the propagation of information from regions (disease foci) to connections (canonical networks). Both of these variables have a straightforward biological meaning and can be used to design follow-up studies.

MULTISITE NETWORK ANALYSIS OF AUTISM

Our primary exploration of ASD relies on the publicly available and multisite autism brain imaging data exchange (ABIDE) (DiMartino et al., 2014). Given the variability of MR acquisition protocols across sites, we focus on four participating institutions, rather than filtering all subjects by some demographic criterion. These sites are the Yale Child Study Center, the Kennedy Krieger Institute, the University of California Los Angeles (Sample 1), and the University of Michigan (Sample 1).

EXPERIMENTAL SETUP

Subject selection: Our study focuses on children and adolescents, 7 to 19 years of age. Inclusion criteria for subjects within the chosen sites were based on both the acquisition quality and successful data preprocessing. On the acquisition front, we required whole-brain coverage and manual inspection of the MPRAGE and BOLD data quality. In addition, we excluded subjects who exhibited significant head motion (>0.5 mm translation or $>0.5°$ rotation) in 25% or more time points of the BOLD series. On the preprocessing side, we verified accurate coregistration between the structural MPRAGE and functional BOLD images. We also filtered individuals for whom the distribution of region-wise rsfMRI correlations was markedly different from all other subjects, as measured by the Hellinger distance. In total, 260 subjects (141 neurotypical, 119 ASD) were selected for analysis. Additional details about the MR acquisition protocols and subject demographics can be found in Di Martino et al. (2014), and Venkataraman et al. (2015).

Data preprocessing: Our Bayesian framework in "An Unbiased Bayesian Framework for Functional Connectomics" section is based on region-wise connectivity measures. Region selection remains an open problem in functional connectomics.

For example, smaller regions are more susceptible to noise artifacts, whereas larger regions can potentially blur the relevant functional effects. This work relies on the Desikan-Killany atlas native to Freesurfer (Fischl et al., 2004), which segments the brain into 86 cortical and subcortical regions that roughly correspond to Broadman areas. The Desikan-Killany atlas provides anatomically meaningful correspondences across subjects that relate to functional divisions in the brain. We emphasize that our method can be applied to any set of consistently defined ROIs across subjects (e.g., the Supplementary Results of (Venkataraman et al., 2015)). The structural ROIs were then projected onto the subject-native fMRI space for each individual.

The BOLD rsfMRI data were processed using Functional MRI of the Brain's Software Library (FSL) (Smith et al., 2004) and in-house matrix laboratory (MATLAB) scripts (MATLAB, 2013). We discarded the first seven rsfMRI time points, and performed motion correction via rigid body alignment and slice timing correction using trilinear/sinc interpolation. The data were spatially smoothed using a Gaussian kernel with 5-mm full width at half maximum (FWHM) and band-pass filtered with cutoffs 0.01 and 0.1 Hz. Next, we regressed global contributions to the time courses from the white matter, ventricles, and whole brain to diminish the influence of physiological noise. Finally, we performed data scrubbing to remove consecutive time points with >0.5 mm translation or >0.5° rotation between them. We computed the observed rsfMRI connectivity measures as the Pearson correlation coefficient between the mean time courses within the two regions. These pairwise connectivity values were then aggregated into an 86×86 rsfMRI data matrix for each subject.

Evaluation criteria: We employed a rigorous evaluation strategy that included both quantitative measures of reproducibility and a qualitative assessment based on the fMRI literature.

Quantitatively, we evaluated the robustness of our approach in two ways: (1) nonparametric permutation tests for statistical significance and (2) bootstrapping experiments to confirm test-retest reliability. The permutation tests allowed us to estimate the null distribution of disease foci. Our procedure was to randomly assign the subject diagnoses (e.g., neurotypical vs ASD) 1000 times, fit the Bayesian model, and compute the region posterior probabilities q_i for each trial. The significance of region i is the proportion of permutations that yield a larger value of q_i for any of the K networks than is obtained under the true labeling. Notice that this is a particularly stringent criterion for $K > 1$, because the previous P-value does not account for interdependencies between the networks. In contrast, the bootstrapping experiment involves fitting the model using a subset of the data while preserving the ratio of neurotypical subjects to ASD patients. We intentionally did not control for other demographic or clinical variables (site, age, IQ, ADOS/ADI scores) to push the limits of our method on heterogeneous data. We resampled the random subsets multiple times and considered the average region posterior probability $\bar{q}_i$ across runs.

On the qualitative side, we leveraged the Neurosynth database (http://www.neurosynth.org/) to provide an unbiased and comprehensive evaluation of the functionality supported by each canonical network. Neurosynth aggregates both the activation coordinates reported in prior fMRI studies and a set of descriptive words/phrases pulled from the abstracts. The metaanalytic framework uses the power of large datasets to calculate the posterior probability $P(Feature|Activation\ Coordinate)$ for a given psychological feature (i.e., word or phrase) at a particular spatial location (Yarkoni et al., 2011). In this way, we can identify constructs that have consistently been associated with a particular activation coordinate across a wide variety of fMRI studies and subject cohorts. At the time of our analysis, Neurosynth had precomputed and stored 3099 brain maps based on the previously discussed posterior information; each map is associated with a particular linguistic feature. Additionally, the website creators have used Latent Dirichlet Allocation (LDA) (Chang et al., 2013) to generate a set of high-level topics from the original 3099 words and phrases.

NETWORK-BASED DIFFERENCES IN ASD

From "An Unbiased Bayesian Framework for Functional Connectomics" section, we observe that the single free parameter of our model is the number of canonical networks K. This parameter can be set based on prior clinical knowledge, or we can sweep its value to track the evolution of canonical network foci with varying model complexity. This book chapter focuses on our result for $K = 2$ networks, which reveals a decoupling between social and language dysfunction in ASD. We refer the reader to our original publication (Venkataraman et al., 2015) and Supplementary Results for a more detailed exposition.

Canonical networks: Fig. 2 (left) illustrates the detected foci (region posterior probability $q_i > 0.5$) for $K = 2$ canonical networks. We have colored each region according to the uncorrected $-\log(P-value)$, such that red indicates low significance and yellow corresponds to high significance. Given the region foci, we can estimate the abnormal functional pathways based on the posterior differences between the neurotypical and ASD functional templates. These connections are shown in Fig. 2 (right) using the BrainNet Viewer toolbox for MATLAB (Xia et al., 2013).

Our primary network consists of four disease foci: the left middle temporal gyrus ($q_i = 0.97$, $P < .001$), the left posterior cingulate ($q_i = 1.00$, $P < .01$), the left supramarginal gyrus ($q_i = 1.00$, $P < .01$), and the right temporal pole ($q_i = 1.00$, $P < .05$). Interestingly, the abnormal pathways indicate a general reduction in long-range connectivity (blue lines) and an overall increase in short-range connectivity (magenta lines) in ASD. Our second network has lower significance and consists of the left banks of the middle superior temporal sulcus ($q_i = 1.00$, $P < .04$), the right posterior superior temporal sulcus extending into inferior parietal lobule ($q_i = 0.86$, $P < .08$), and the right middle temporal gyrus ($q_i = 0.98$, $P < .07$). We accepted a lower significance threshold for this network due to our stringent criteria of computing p-values for $K > 1$. The corresponding functional pathways demonstrate reduced interhemispheric connectivity but largely increased intrahemispheric connectivity.

Model robustness: Fig. 3 reports the average posterior probability in our test-retest experiments. We use half of the subjects in each trial while preserving the ratio of ASD patients to neurotypical controls. We have displayed only the regions for

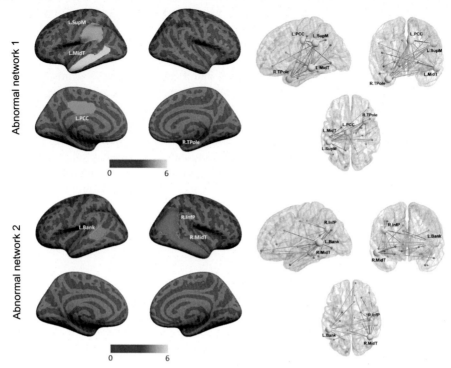

FIG. 2

Canonical networks inferred by our Bayesian model for $K = 2$ using the ABIDE dataset. Left: Significant regions based on permutation tests (region posterior probability $qi > 0.5$, uncorrected $P < .08$). Significance is computed via the likelihood of a region appearing in *either* network. The *color bar* corresponds to the negative log P-value. Right: Estimated graphs of abnormal functional connectivity in ASD. The *yellow nodes* correspond to disease foci. *Blue lines* indicated reduced functional connectivity in ASD, and *magenta lines* denote increased functional connectivity in ASD.

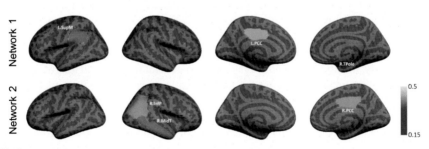

FIG. 3

Average marginal posterior probability qi for each network across 50 random samplings from the ABIDE dataset. Each subset includes 50% of the subjects, such that the ratio of ASD patients to neurotypical controls is preserved. The *color bar* denotes the average posterior probability. The highlighted regions correspond to the supramarginal gyrus (L.SupM), the inferior parietal cortex (R.InfP), the middle temporal gyrus (R.MidT), and the posterior cingulate (L.PCC and R.PCC).

which the average posterior probability across 50 random trials is >0.15—thereby emphasizing only the most prominent patterns. The color bar indicates the average probability, such that yellow denotes the strongest foci and red corresponds to the weakest influence. Remarkably, despite using only half of the data, our test-retest experiments are able to recover many of the disease foci from Fig. 2. This result verifies the generalizability of our framework for localizing robust functional connectivity differences in ASD.

Metaanalysis of the fMRI Literature: Fig. 4 illustrates our Neurosynth results. The upper panel describes the top 10 LDA topics implied by each set of network foci. We omitted topics that describe brain anatomy (e.g., default mode) or a neurological disorder (e.g., autism, which was among the top 10 for both networks). These mental states reveal both overlap and clear functional distinctions between the two intrinsic networks. Network 1 was associated with language-related topics, including comprehension and semantic processing. Network 2 also loaded heavily on language constructs but was uniquely associated with social-related topics, such as person and self-referential processing. The bottom panels of Fig. 4 display the relative correlation strengths of each topic generated by the Neurosynth decoder.

Implications for ASD: Language and communication deficits are among the defining features of ASD, as supported by our highly significant canonical network 1. Task-based fMRI has shed light on the system-level organization of language processing. For example, the upper band of the STS responds preferentially to the human voice, in comparison to other acoustic signals (Boddaert et al., 2003; Hickok, 2009). Going one step further, perception of meaningful speech localizes to the middle and inferior temporal cortices, whereas sentence comprehension tasks activate the bilateral superior temporal gyri (Price, 2009). On the other hand, canonical network 2 pinpoints a well-known social perception pathway centered in the right posterior STS, extending into the inferior parietal lobule and right middle temporal gyrus (Yang et al., 2015). The posterior STS is sensitive to and selective for social stimuli that signal intent in humans (Jastorff et al., 2012). In the visual domain, it activates preferentially to faces versus objects and to socially meaningful human actions versus nongoal-directed movements (Bahnemann et al., 2010; Gobbini and Haxby, 2007; Watson et al., 2014). In the auditory domain, the posterior STS responds to auditory speech (Ethofer et al., 2006; Wildgruber et al., 2006). Finally, the posterior STS is functionally interconnected to all the key regions of the "social brain" (Yang et al., 2015).

In addition to the region foci, the altered functional pathways are highly relevant to current theories of ASD. Specifically, our canonical networks support a general reduction in long-range connectivity, both within and across the hemispheres. These long-range connections likely correspond to integration processes, which are essential to higher order social, emotional, and communicative functioning (Oberman and Ramachandran, 2008; Rippon et al., 2007; Wolff et al., 2012). It is believed that disruptions in long-range connectivity play a key role in the hallmark dysfunctions of ASD. Likewise, hemispheric abnormalities are prevalent in the ASD literature, particularly within the language domain (Dawson et al., 1989; Redcay and Courchesne, 2008).

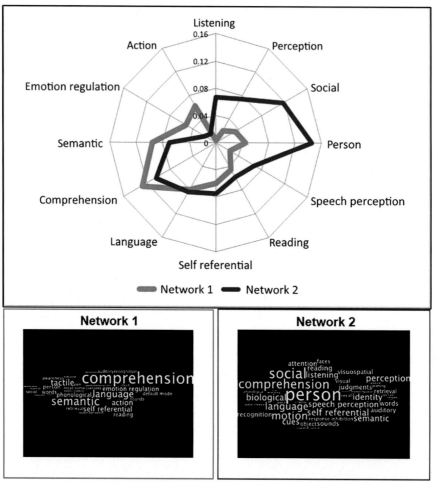

FIG. 4

Upper: correlation values of the top 10 topics for each network, which represent the specificity of neurocognitive functions derived from metaanalytic decoding. We include words that overlap between networks, resulting in 14 total features. Lower: relative rankings of the psychological constructs. Word size reflects rank-ordered correlation coefficients, which emphasizes terms most associated with each network (Poldrack et al., 2009).

In contrast, both canonical networks also find a general increase in short-range connectivity, which may contribute to enhanced cognitive and pattern recognition skills often reported in high-functioning ASD subpopulations (Johnson et al., 2002).

It is worth mentioning that our framework does not implicate certain regions that have been previously reported in the ASD literature. Examples include the prefrontal cortex, as related to working memory and executive function (Baron-Cohen et al., 1999;

Courchesne and Pierce, 2005; Gilbert et al., 2008; Just et al., 2004), and the visuomotor cortex (Dowell et al., 2009; Koshino et al., 2008; Mostofsky and Ewen, 2011; Nebel et al., 2016). One possibility is that such differences are overwhelmed by the intersite variability, including scanner model, acquisition procedures, and subject recruitment. Another possibility is that there is weak evidence in our cohort to support the prefrontal and visuomotor cortices acting as foci. Recall that our model is completely data-driven and does not impose spatial constraints on the canonical networks. Finally, it is possible that our region parcellation is too coarse to identify these additional effects.

We conclude this section by addressing the model complexity parameter K, which specifies the number of canonical networks that explain the data. Intuitively, there will be a tradeoff between model robustness and our ability to pinpoint weaker effects. Due to the lack of ground-truth connectivity information, we have constructed both quantitative and qualitative proxies for generalizability and relevance, which can be used to evaluate the impact of K. Taken together, the results in this chapter demonstrate the power of our hierarchical Bayesian framework as a future analysis tool for ASD.

TOWARD CHARACTERIZING PATIENT HETEROGENEITY

As described earlier in this chapter, ASD is a notoriously heterogeneous disorder that includes both a wide variety of behavioral symptoms and a range of clinical severity. As previously alluded to, the existing ASD literature implicitly treats the patient group as homogenous, for example, by conducting a statistical evaluation that separates patients from controls. Even our unbiased Bayesian framework infers a single functional template for the entire ASD cohort. It has been argued that this gross simplification is partially responsible for the lack of reproducible fMRI findings in much of the clinical neuroscience literature (Horga et al., 2014).

This section tackles a fundamental yet commonly overlooked question in the study of functional connectomics: how do we identify the altered functional pathways given a heterogeneous patient cohort? As a first step, we consider a one-dimensional measure of clinical severity and a single canonical network ($K = 1$). We fold the behavioral information into our probabilistic framework by stipulating that the canonical network influence on an individual patient is moderated by their scalar severity score. Said another way, patients with higher ASD severity scores will manifest *more* network dysfunction than patients with lower ASD severity. Hence, rather than dismissing or regressing out the clinical scores, these measures will crucially guide our network estimation procedures.

Fig. 5 illustrates the revised interactions. As seen, the strengths of the canonical network edges are proportional to behavioral score $\beta_m \in [0, 1]$, which can be quantified via neuropsychiatric testing or parent questionnaires. Effectively, the patient likelihood weighs the relative contributions of the clinical and neurotypical templates via β_m. Mathematically, the patient rsfMRI correlation $\bar{B}_{ij}^m$ is sampled from

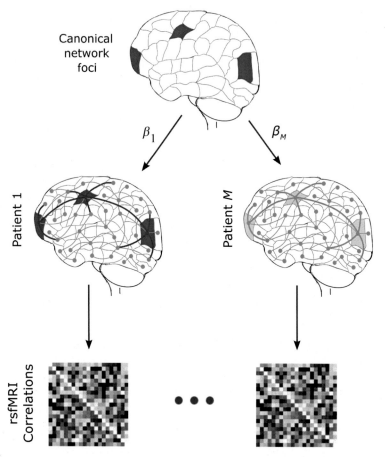

Canonical
network
foci

β_1 β_M

Patient 1 Patient M

rsfMRI
Correlations

FIG. 5

Conceptual diagram of behavioral influence. *Red regions* correspond to the disease foci, and *red edges* specify the canonical functional network. *Green edges* are normal (i.e., healthy) connections. The canonical network contribution for each patient m is specified by the clinical severity, $\beta m \in [0, 1]$. Here, $\beta_1 > \beta M$, as indicated by the *darker edges*.

the Gaussian distribution specified by the ASD template with probability β_m, and it is sampled from the Gaussian distribution associated with the neurotypical template with probability $1 - \beta_m$.

EXPERIMENTAL SETUP

Study participants: To avoid site artifacts, we rely on a much tighter sample drawn from the Kennedy Krieger Institute in Baltimore, MD. Our dataset consists of 66 high-functioning, school-aged children with ASD and 66 neurotypical controls, who were matched on the basis of age, gender, and IQ. The severity measures β_m in our analysis correspond to the Autism Diagnostic Observation Schedule (ADOS) total raw score (Gotham et al., 2007), normalized by the maximum possible test score.

Data preprocessing: Once again, we must define consistent region boundaries for our model. In this section, we rely on the Automatic Anatomical Labeling (AAL) atlas (Tzourio-Mazoyer et al., 2002) to delineate 116 cortical, subcortical, and cerebellar regions. RsfMRI preprocessing was implemented using the Statistical Parametric Mapping (SPM) toolbox, in-house MATLAB scripts, and the Analysis of Functional NeuroImages (AFNI) package (Cox, 1996). Our basic fMRI pipeline included slice timing correction, rigid body realignment, and normalization to the EPI version of the MNI template. To facilitate the connectivity analyses, we temporally detrended the time series and used CompCorr to estimate and remove spatially coherent noise from the white matter, ventricles, motion parameters, and their first derivatives (Behzadi et al., 2007). The cleaned data was spatially smoothed with a 6-mm FWHM Gaussian kernel, temporally filtered and spike-corrected. Similar to the previous section, our rsfMRI measures are computed as the Pearson correlation coefficients between the mean time courses in the two regions. We focus on deviations from baseline synchrony by centering the whole-brain correlation histograms for each subject.

Evaluation criteria: Similar to the previous section, we evaluated the test-retest reliability of our region assignments q_i via bootstrapping. Specifically, we fitted the heterogeneous model to random subsets of the data while preserving the ratio of ASD patients to neurotypical controls. We ran two analyses, which corresponded to subsets with 90% and 50% of the overall cohort. Our quantitative measure of robustness is the average region posterior probability across 100 random trials. In addition to bootstrapping, we performed a qualitative comparison of this revised Bayesian model with our original framework in "An Unbiased Bayesian Framework for Functional Connectomics" section and with univariate *t*-tests on the pairwise rsfMRI correlation values.

NETWORK DYSFUNCTION LINKED TO ASD SEVERITY

Heterogeneous network architecture: Fig. 6 illustrates the single canonical network inferred by our model. The yellow nodes correspond to the disease foci, and we have displayed connections consistently implicated across bootstrapping trials. Similar to the previous section, the magenta and blue lines denote increased and reduced latent connectivity in ASD relative to the neurotypical population. As seen, our model identifies four disease foci: the right precentral gyrus (R.PreCG), the right posterior cingulate gyrus (R.PCG), the right angular gyrus (R.ANG), and vermis 8 of the cerebellum (Verm8).

Our results are closely aligned with recent findings in ASD. For example, the right precentral gyrus and cerebellar vermis represent foci specialized in the production of actions, a core behavioral feature consistently reported in ASD

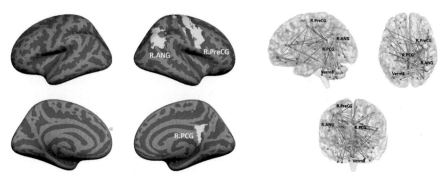

FIG. 6

Results of our heterogeneous patient model. Left: disease foci projected onto the inflated cortical surface. Right: canonical network of abnormal functional connectivity. *Yellow nodes* correspond to the disease foci. *Blue lines* signify reduced functional connectivity in ASD; *magenta lines* denote increased functional connectivity in ASD.

(Mostofsky and Ewen, 2011; Nebel et al., 2016). Moreover, the right angular gyrus and posterior cingulate are common to the DMN, which is believed to moderate internal reflective processes (Buckner et al., 2008). Converging multimodal evidence suggests that the altered functional and structural organization of the DMN contributes to social cognitive dysfunction in ASD (Padmanabhana et al., 2017). In this manner, our canonical network seems to represent the consequence of abnormal development and compensatory mechanisms for the expression of social and communicative functions.

Model robustness: Fig. 7 reports the average posterior probability $\bar{q}_i$ of each region across 100 bootstrapped trials. We display only the regions for which $\bar{q}_i > 0.3$ to emphasize the most prominent patterns. Notice that our model consistently recovers

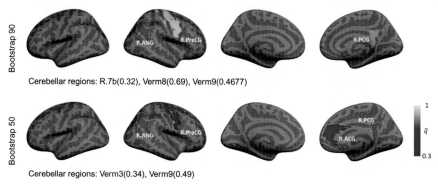

FIG. 7

Average marginal posterior probability $\bar{q}_i$ for each community across 100 random samplings of the rsfMRI dataset. Top row includes 90% of the subjects in each subset, and the bottom row includes 50%. Reproducibility of cerebellar regions are listed underneath.

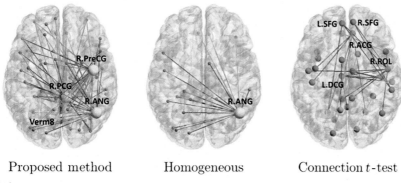

Proposed method   Homogeneous   Connection t-test

FIG. 8

Qualitative comparison of our heterogeneous patient model (left), the original Bayesian model described in "An Unbiased Bayesian Framework for Functional Connectomics" section, and the top connections ($P < .001$ uncorrected) identified via two-sample t-tests on the pairwise correlation values (right).

the canonical network foci in Fig. 6 when trained on 90% of the data. Remarkably, we are still able to detect the original network foci using just half the dataset, which further validates the reproducibility of our revised Bayesian framework. In fact, we note that the robustness in Fig. 7 is markedly higher than in our initial ABIDE study, in part due to stricter control over the subject recruitment and MR acquisition protocols. Finally, our bootstrapping experiments also implicate cerebellar regions adjacent to Vermis 8, which ties into broader theories of altered cerebellar functioning in ASD (Becker and Stoodley, 2013).

Baseline comparisons: Fig. 8 compares our heterogeneous canonical network (left) with our original Bayesian model (middle), which assumes a homogeneous patient group, and with standard univariate tests (right). Notice that the homogeneous model identifies a single disease focus in the DMN (R.ANG). However, incorporating the severity scores β_m seems to provide an additional level of flexibility, which allows us to find more subtle effects in other brain regions. On the other hand, connections implicated by two-sample t-tests form a markedly different pattern than our network model results. First, the univariate tests implicate several isolated connections, which are difficult to interpret. Second, the connections tend to concentrate in the frontal cortex and anterior cingulate gyrus, rather than the DMN. This observation suggests that our disease foci are unveiling a unique facet of the rsfMRI data.

CONCLUDING REMARKS

Functional connectomics has become a universal tool to noninvasively assess neural interactions on a global scale. Going one step further, it allows us to isolate system-level dysfunctions induced by a complex neurological disorder, such as ASD. Recent findings have confirmed a general reduction in long-range connectivity between functional systems, which may be linked to integrative processes, such as social awareness and emotional perception. Graph theoretic studies suggest that the autistic brain might be more "randomly" organized than a neurotypical brain. However, despite the wealth of research and publications, we have yet to identify robust rsfMRI biomarkers for the disorder. This paradox can be explained by one of two possibilities: either the null hypothesis is true and rsfMRI does not contain enough information to be clinically relevant for ASD, or we have not sufficiently explored the methods that can describe the functional abnormalities, while handling the quirks of rsfMRI data.

In an effort to address this conundrum, we have devoted the bulk of this chapter to a novel probabilistic framework that contributes two unexplored dimensions to current network analytics: (1) we assume that the *functional differences* attributed to ASD form their own subnetwork, and (2) we explicitly model clinical severity. The corresponding network architectures confirm theories of both impaired social communication and reduced sensorimotor integration in ASD. Our results are quantitatively verified via nonparametric permutation testing and test-retest reliability experiments. Equally exciting, our Bayesian model can be adapted to a variety of technical challenges from detecting abnormal patterns in task fMRI (Venkataraman et al., 2016a, b), evaluating pre- versus posttreatment connectivity differences (Venkataraman et al., 2016a, b), multimodal integration with diffusion MRI (Venkataraman et al., 2013b), and a patient-specific analysis (Sweet et al., 2013). On the clinical side, our framework has been applied to ASD, schizophrenia, posttraumatic stress disorder, mild traumatic brain injury, and epilepsy.

Looking forward, there are a number of challenges that we, as a field, must address with respect to the functional connectomics of ASD. One obvious factor is patient heterogeneity. "Toward Characterizing Patient Heterogeneity" section takes a first step of incorporating clinical severity into the network decomposition. However, ASD is characterized by a multitude of behavioral and cognitive symptoms, from impaired social skills to language problems to stereotyped behaviors. To complicate matters, full-scale intelligence also plays a role in functional connectivity. Future methods should leverage this variability to provide a more complete picture of the disorder. A second challenge is to develop methods that are robust to data acquisition and image quality, while providing interpretable information about ASD. The issue of interpretability will become increasingly relevant in the next few years given the rise of artificial intelligence. In particular, while neural networks and deep learning can achieve impressive detection and regression performance, they are essentially black-box functions, which provide little insight as to what patterns in the data are meaningful and why. Along the same lines, a third goal is to formalize new evaluation metrics. At present, statistical significance dominates the ASD literature, but p-values are often poor indicators of test-retest reliability (Venkataraman et al., 2010). Finally, we close this chapter on a philosophical note. Functional connectomics has been invaluable to studying and probing complex questions related to ASD. We have found, albeit contradictory, patterns related to higher level cognitive functions, such as social awareness and sensory integration. But what comes next? How do we translate these findings into a real-world impact at the patient level? These questions should fuel the next great wave of autism research.

REFERENCES

Achard, S., Bullmore, E., 2007. Efficiency and cost of economical brain functional networks. PLoS Comput. Biol. 3 (2), 1–10.

Americal Psychiatric Association, 2013. Diagnostic and Statistical Manual of Mental Disorders, fifth ed. American Psychiatric Publishing, Arlington, VA.

Aoki, Y., Abe, O., Nippashi, Y., Yamasue, H., 2013. Comparison of white matter integrity between autism spectrum disorder subjects and typically developing individuals: a meta-analysis of diffusion tensor imaging Tractography studies. Mol. Autism 4 (1), 25.

Assaf, M., Jagannathan, K., Calhoun, V.D., Miller, L., Stevens, M.C., Sahl, R., 2010. Abnormal functional connectivity of default mode sub-networks in autism spectrum disorder patients. Neuroimage 53, https://doi.org/10.1016/j.neuroimage.2010.05.067.

Bahnemann, M., Dziobek, I., Prehn, K., Wolf, I., Heekeren, H.R., 2010. Sociotopy in the temporoparietal cortex: common versus distinct processes. Soc. Cogn. Affect. Neurosci. 5 (1), 48–58. https://doi.org/10.1093/scan/nsp045.

Baron-Cohen, S., Ring, H.A., Wheelwright, S., Bullmore, E.T., Brammer, M.J., Simmons, A., Williams, S.C.R., 1999. Social intelligence in the normal and autistic brain: an fMRI study. Eur. J. Neurosci. 11 (6), 1891–1898. https://doi.org/10.1046/j.1460-9568.1999.00621.x.

Bassett, D.S., Bullmore, E., 2006. Small-world brain networks. Neuroscientist 12, 512–523.

Bassett, D.S., Bullmore, E., 2009. Human brain networks in health and disease. Curr. Opin. Neurol. 22, 340–347.

Becker, E., Stoodley, C., 2013. Autism spectrum disorder and the cerebellum. Int. Rev. Neurobiol. 113, 1–34.

Behzadi, Y., Restom, K., Liau, J., Liu, T.T., 2007. A component based noise correction method (CompCor) for BOLD and perfusion based fMRI. Neuroimage 37 (1), 90–101.

Bell, A.J., Sejnowski, T.J., 1995. An information-maximization approach to blind separation and blind deconvolution. Neural Comput. 7, 1129–1159.

Belmonte, M.K., Yurgelun-Todd, D.A., 2003. Functional anatomy of impaired selective attention and compensatory processing in autism. Cogn. Brain Res. 17 (3), 651–664. https://doi.org/10.1016/S0926-6410(03)00189-7.

Boddaert, N., Belin, P., Chabane, N., Poline, J.B., Barthélémy, C., Mouren-Simeoni, M.C., Zilbovicius, M., 2003. Perception of complex sounds: abnormal pattern of cortical activation in autism. Am. J. Psychiatry 160 (11), 2057–2060. Retrieved from, http://view.ncbi.nlm.nih.gov/pubmed/14594758.

Buckner, R.L., Vincent, J.L., 2007. Unrest at rest: default activity and spontaneous network correlations. Neuroimage 37 (4), 1091–1096.

Buckner, R.L., Andrews-Hanna, J.R., Schacter, D.L., 2008. The Brain's default network anatomy, function, and relevence to disease. Ann. NY Acad. Sci. 1124, 1–38.

Bullmore, E., Sporns, O., 2009. Complex brain networks: graph theoretical analysis of structural and functional systems. Nat. Rev. Neurosci. 10, 186–198.

Calhoun, V.D., Adali, T., Hansen, L.K., Larsen, J., Pekar, J.J., 2003. In: *ICA of Functional MRI Data: An Overview. The 4th International Symposium on Independent Component Analysis and Blind Signal Separation*, pp. 281–288.

Calhoun, V.D., Liu, J., Adalı, T., 2009. A review of group ICA for fMRI data and ICA for joint inference of imaging, genetic, and ERP data. Neuroimage 45, S163–S172.

Centers for for Disease Control, 2012. Prevalence of autism spectrum disorders—autism and developmental disabilities monitoring network, 14 sites, United States, 2008. MMWR 61 (3), 1–19.

Cerliani, L., Mennes, M., Thomas, R.M., Di Martino, A., Thioux, M., Keysers, C., 2015. Increased functional connectivity between subcortical and cortical resting-state networks in autism Spectrum disorder. JAMA Psychiat. 72 (8), 767–777. https://doi.org/10.1001/jamapsychiatry.2015.0101.

Chang, L.J., Yarkoni, T., Khaw, M.W., Sanfey, A.G., 2013. Decoding the role of the insula in human cognition: functional parcellation and large-scale reverse inference. Cereb. Cortex 23 (3), 739–749. https://doi.org/10.1093/cercor/bhs065.

Cherkassky, V.L., Kana, R.K., Keller, T.A., Just, M.A., 2006. Functional connectivity in a baseline resting-state network in autism. Neuroreport 17 (16), 1687–1690.

Courchesne, E., Pierce, K., 2005. Why the frontal cortex in autism might be talking only to itself: Local over-connectivity but long-distance disconnection. Curr. Opin. Neurobiol. 15, 225–230.

Cox, R.W., 1996. AFNI: software for analysis and visualization of functional magnetic resonance neuroimages. Comput. Biomed. Res. 29 (3), 162–173.

Dawson, G., Finley, C., Phillips, S., Lewy, A., 1989. A comparison of hemispheric asymmetries in speech related brain potentials of autistic and dysphasic children. Brain Lang. 37 (1), 26–41.

Delmonte, S., Gallagher, L., O'Hanlon, E., McGrath, J., Balsters, J.H., 2013. Functional and structural connectivity of frontostriatal circuitry in autism spectrum disorder. Front. Hum. Neurosci. 7, 430. https://doi.org/10.3389/fnhum.2013.00430.

Di Martino, A., Yan, C.-G., Li, Q., Denio, E., Castellanos, F.X., Alaerts, K., Milham, M.P., 2014. The autism brain imaging data exchange: towards a large-scale evaluation of the intrinsic brain architecture in autism. Mol. Psychiatry 19, 659–667.

DiMartino, A., Yan, C.G., Li, Q., Denio, E., Castellanos, F.X., Alaerts, K., et al., 2014. The autism brain imaging data exchange: towards a large-scale evaluation of the intrinsic brain architecture in autism. Mol. Psychiatry 19 (6), https://doi.org/10.1038/mp.2013.78.

Dowell, L.R., Mahone, M.E., Mostofsky, S.H., 2009. Associations of postural knowledge and basic motor skill with dyspraxia in autism: Implication for abnormalities in distributed connectivity and motor learning. Neuropsychology 23 (5), 563–570.

Estrada, E., Hatano, N., 2008. Communicability in complex networks. Phys. Rev. Stat. 77, 036111.

Ethofer, T., Anders, S., Erb, M., Herbert, C., Wiethoff, S., Kissler, J., Crodd, W., Wildgruber, D., 2006. Cerebral pathways in processing of affective prosody: a dynamic causal modeling study. Neuroimage 30 (2), 580–587. https://doi.org/10.1016/j.neuroimage.2005.09.059.

Fischl, B., Salat, D.H., van der Kouwe, A.J.W., Makris, N., Ségonne, F., Quinn, B.T., Dale, A.M., 2004. Sequence-independent segmentation of magnetic resonance images. Neuroimage 23, 69–84.

Fox, M.D., Raichle, M.E., 2007. Spontaneous fluctuations in brain activity observed with functional magnetic resonance imaging. Nature 8, 700–711.

Gabrieli-Whitfield, S., Thermenos, H.W., Milanovic, Z., Tsuang, M.T., Faraone, S.V., McCarley, R.W., Shenton, M.E., Green, A.I., Nieto-Castanon, A., La Violette, P., Wojcik, J., Gabrieli, J.D., Seidman, L.J., 2009. Hyperactivity and Hyperconnectivity of the default network in schizophrenia and in first-degree relatives of persons with schizophrenia. Natl. Acad. Sci. USA 106, 1279–1284.

Geschwind, D.H., Konopka, G., 2009. Neuroscience in the era of functional genomics and systems biology. Nature 461 (7266), 908–915.

Geschwind, D.H., Levitt, P., 2007. Autism spectrum disorders: developmental disconnection syndromes. Curr. Opin. Neurobiol. 17, 103–111.

Gilbert, S.J., Bird, G., Brindley, R., Frith, C.D., Burgess, P.W., 2008. A typical recruitment of medial prefrontal cortex in autism spectrum disorders: an fMRI study of two executive function tasks. Neuropsychologia 46 (9), 2281–2291. https://doi.org/10.1016/j.neuropsychologia.2008.03.025.

Globerson, E., Amir, N., Kishon-Rabin, L., Golan, O., 2015. Prosody recognition in adults with high-functioning autism Spectrum disorders: from psychoacoustics to cognition. Autism Res. 8 (2), 153–163.

Gobbini, M.I., Haxby, J.V., 2007. Neural systems for recognition of familiar faces. Neuropsychologia 45 (1), 32–41. Retrieved from, http://view.ncbi.nlm.nih.gov/pubmed/16797608.

Gotham, K., Risi, S., Pickles, A., Lord, C., 2007. The autism diagnostic observation schedule: revised algorithms for improved diagnostic validity. J. Autism Dev. Disord. 37, 613–627.

Grossman, R.B., Bemis, R.H., Plesa Skwerer, D., Tager-Flusberg, H., 2010. Lexical and affective prosody in children with high-functioning autism. J. Speech Lang. Hear. Res. 53 (3), 778–793.

Hernandez, L.M., Rudie, J.D., Green, S.A., Bookheimer, S., Dapretto, M., 2015. Neural signatures of autism spectrum disorders: insights into brain network dynamics. Neuropsychopharmacol. Rev. 40 (1), 171–189.

Hickok, G., 2009. The functional neuroanatomy of language. Phys. Life Rev. 6 (3), 121–143. https://doi.org/10.1016/j.plrev.2009.06.001.

Honey, C.J., Sporns, O., Cammoun, L., Gigandet, X., Thiran, J.P., Meuli, R., Hagmann, P., 2009. Predicting human resting-state functional connectivity from structural connectivity. Proc. Natl. Acad. Sci. 106, 2035–2040.

Horga, G., Kaur, T., Peterson, B.S., 2014. Annual research review: Current limitations and future directions in {MRI} studies of child- and adult-onset developmental psychopathologies. J. Child Psychol. Psychiatry 55 (6), 659–680.

Hull, J.V., Jacokes, Z.J., Torgerson, C.M., Irimia, A., Van Horn, J.D., 2016. Resting-state functional connectivity in autism Spectrum disorders: a review. Front. Psych. 7, 205. https://doi.org/10.3389/fpsyt.2016.00205.

Itahashi, T., Yamada, T., Watanabe, H., Nakamura, M., Jimbo, D., Shioda, S., Hashimoto, R., 2014. Altered network topologies and hub organization in adults with autism: a resting-state fMRI study. PLoS One 9 (4), e94115.

Itahashi, T., Yamada, T., Watanabe, H., Nakamura, M., Ohta, H., Kanai, C., Hashimoto, R., 2015. Alterations of local spontaneous brain activity and connectivity in adults with high-functioning autism Spectrum disorder. Mol. Autism 6 (1), 30. https://doi.org/10.1186/s13229-015-0026-z.

Jastorff, J., Popivanov, I.D., Vogels, R., Vanduffel, W., Orban, G.A., 2012. Integration of shape and motion cues in biological motion processing in the monkey STS. Neuroimage 60 (2), 911–921. https://doi.org/10.1016/j.neuroimage.2011.12.087.

Johnson, M.H., Halit, H., Grice, S.J., Karmiloff-Smith, A., 2002. Neuroimaging of typical and atypical development: a perspective from multiple levels of analysis. Dev. Psychopathol. 41, 521–536.

Jordan, M.I., Ghahramani, Z., Jaakkola, T.S., Saul, L.K., 1999. An introduction to variational methods for graphical models. Mach. Learn. 37, 183–233.

Just, M.A., Cherkassky, V.L., Keller, T.A., Minshew, N.J., 2004. Cortical activation and synchronization during sentence comprehension in high-functioning autism: evidence of underconnectivity. Brain 127, 1811–1821.

Just, M.A., Keller, T.A., Malavea, V.L., Kana, R.K., Varmac, S., 2012. Autism as a neural systems disorder: a theory of frontal-posterior underconnectivity. Neurosci. Biobehav. Rev. 36, 1292–1313.

Kanner, L., 1943. Autistic disturbances of affective contact. Neurodiver. Child. 2, 217–250.

Kennedy, D.P., Courchesne, E., 2008a. Functional abnormalities of the default network during self-and other-reflection in autism. Soc. Cogn. Affect. Neurosci. 3, 177–190.

Kennedy, D.P., Courchesne, E., 2008b. The intrinsic functional organization of the brain is altered in autism. Neuroimage 39, 1877–1887.

Keown, C.L., Shih, P., Nair, A., Peterson, N., Mulvey, M.E., Muller, R.A., 2013. Local functional overconnectivity in posterior brain regions is associated with symptom severity in autism spectrum disorders. Cell Rep. 5, 567–572.

Kleinhans, N.M., Müller, R.A., Cohen, D.N., Courchesne, E., 2008. Atypical functional lateralization of language in autism spectrum disorders. Brain Res. 1221, 115–125. https://doi.org/10.1016/j.brainres.2008.04.080.

Koshino, H., Carpenter, P.A., Minshew, N.J., Cherkassky, V.L., Keller, T.A., Just, M.A., 2005. Functional connectivity in an fMRI working memory task in high-functioning autism. Neuroimage 24 (3), 810–821.

Koshino, H., Kana, R.K., Keller, T.A., Cherkassky, V.L., Minshew, N.J., Just, M.A., 2008. fMRI investigation of working memory for faces in autism: visual coding and underconnectivity with frontal areas. Cereb. Cortex 18 (2), 289–300.

Leslie, D.L., Martin, A., 2007. Health care expenditures associated with autism spectrum disorders. Arch. Pediatr. Adolesc. Med. 161 (4), 350–355.

Liang, M., Zhou, Y., Jiang, T., Liu, Z., Tian, L., Liu, H., Hao, Y., 2006. Widespread functional disconnectivity in schizophrenia with resting-state functional magnetic resonance imaging. Neuro Rep Brain Imag 17 (2), 209–213.

MATLAB, 2013. Version 8.2.0.701 (R2013b). The MathWorks Inc, Natick, MA.

McKeown, M.J., Makeig, S., Brown, G.G., Jung, T.-P., Kindermann, S.S., Bell, A.J., Sejnowski, T.J., 1998. Analysis of fMRI data by blind separation into spatial independent components. Hum. Brain Mapp. 6, 160–188.

Melillo, R., Leisman, G., 2011. Autistic spectrum disorders as functional disconnection syndrome. Rev. Neurosci. 20 (2), 111–131.

Meunier, D., Lambiotte, R., Fornito, A., Ersche, K.D., Bullmore, E.T., 2009. Hierarchical modularity in human brain functional networks. Front. Neuroinform. 3, 37.

Mostofsky, S.H., Ewen, J.B., 2011. Altered connectivity and action model formation in autism. Neuroscientist 17, 437–448.

Nebel, M.B., et al., 2014. Precentral gyrus functional connectivity signatures of autism. Front. Syst. Neurosci. 8, 1–11.

Nebel, M.B., Eloyan, A., Nettles, C.A., Sweeney, K.L., Ament, K., Ward, R.E., Choe, A.S., Barber, A.D., Pekar, J.J., Mostofsky, S.H., 2016. Intrinsic visual-motor synchrony correlates with social deficits in autism. Biol. Psychiatry 79, 633–641.

Oberman, L.M., Ramachandran, V.S., 2008. Preliminary evidence for deficits in multisensory integration in autism spectrum disorders: The mirror neuron hypothesis. Soc. Neurosci. 3, 348–355.

Padmanabhana, A., Lynchd, C.J., Schaere, M., Menon, V., 2017. The default mode network in autism. Biol. Psych. Cogn. Neurosci. Neuroimag. 2 (6), 476–486.

Pelphrey, K.A., Yang, D.-J., McPartland, J.C., 2014. Building a social neuroscience of autism spectrum disorder. Curr. Top. Behav. Neurosci. 16, 215–233.

Poldrack, R.A., Halchenko, Y.O., Hanson, S.J., 2009. Decoding the large-scale structure of brain function by classifying mental states across individuals. Psychol. Sci. 20 (11), 1364–1372.

Price, C., 2009. The anatomy of language: A review of 100 fMRI studies published in 2009. Ann. N. Y. Acad. Sci. 1191, 62–88. https://doi.org/10.1111/j.1749-6632.2010.05444.x.

Redcay, E., Courchesne, E., 2008. Deviant functional magnetic resonance imaging patterns of brain activity to speech in 2-3-year-old children with autism spectrum disorder. Biol. Psychiatry 64 (7), 589–598. https://doi.org/10.1016/j.biopsych.2008.05.020.

Redcay, E., Moran, J.M., Mavros, P.L., Tager-Flusberg, H., Gabrieli, J.D.E., Whitfield-Gabrieli, S., 2013. Intrinsic functional network organization in high-functioning adolescents with autism spectrum disorder. Front. Hum. Neurosci. 7, 573. https://doi.org/10.3389/fnhum.2013.00573.

Rippon, G., Brock, J., Brown, C., Boucher, J., 2007. Disordered connectivity in the autistic brain: challenges for the "new psychophysiology". Int. J. Psychophysiol. 63, 164–172.

Rubinov, M., Sporns, O., 2010. Complex network measures of brain connectivity: uses and interpretations. Neuroimage 52, 1059–1069.

Rudie, J.D., Dapretto, M., 2017. Convergent evidence of brain overconnectivity in children with autism? Cell Rep. 5 (3), 565–566. https://doi.org/10.1016/j.celrep.2013.10.043.

Schmitz, N., Rubia, K., van Amelsvoort, T., Daly, E., Smith, A., and Murphy, D. G. M. (2008). Neural correlates of reward in autism. *The* Br. J. Psychiatry, 192(1), 19-24. Retrieved from http://bjp.rcpsych.org/content/192/1/19.abstract

Scott-Van Zeeland, A.A., Dapretto, M., Ghahremani, D.G., Poldrack, R.A., Bookheimer, S.Y., 2010. Reward processing in autism. Autism Res. 3 (2), 53–67. https://doi.org/10.1002/aur.122.

Smith, S.M., 2012. The future of fMRI connectivity. Neuroimage 62, 1257–1266.

Smith, S.M., Jenkinson, M., Woolrich, M.W., Beckmann, C.F., Behrens, T.E.J., Johansen-Bern, H., Bannister, P.R., De Luca, M., Drobnjak, I., Flitney, D.E., Niazy, R.K., Saunders, J., Vickers, J., Zhang, Y., De Stefano, N., Brady, J.M., Matthews, P.M., 2004. Advances in functional and structural MR image analysis and implementation as FSL. Neuroimage 23 (S1), 208–219.

Sporns, O., Betzel, R.F., 2016. Modular brain networks. Annu. Rev. Psychol. 672, 613–640.

Starck, T., Nikkinen, J., Rahko, J., Remes, J., Hurtig, T., Haapsamo, H., Jussila, K., Kuusikko-Gauffin, S., Mattila, M.L., Jansson-Verkasalo, E., Pauls, D.L., Ebeling, H., Moilanen, I., Tervonen, O., Kiviniemi, V.J., 2013. Resting state fMRI reveals a default mode dissociation between retrosplenial and medial prefrontal subnetworks in ASD despite motion scrubbing. Front. Hum. Neurosci. 7, 802. https://doi.org/10.3389/fnhum.2013.00802.

Stuart, M., McGrew, J.H., 2009. Caregiver burden after receiving a diagnosis of an autism spectrum disorder. Res. Autism Spect. Dis. 3 (1), 86–97.

Stufflebeam, S.M., Liu, H., Sepulcre, J., Tanaka, N., Buckner, R.L., Madsen, J.R., 2011. Localization of focal epileptic discharges using functional connectivity magnetic resonance imaging. J. Neurosurg. 114 (6), 1693–1697.

Sullivan, K., Stone, W.L., Dawson, G., 2014. Potential neural mechanisms underlying the effectiveness of early intervention for children with autism spectrum disorder. Res. Dev. Disabil. 35, 2921–2932.

Supekar, K., Uddin, L.Q., Prater, K., Amin, H., Greicius, M.D., Menon, V., 2010. Development of functional and structural connectivity within the default mode network in young children. Neuroimage 52 (1), 290–301. https://doi.org/10.1016/j.neuroimage.2010.04.009.

Sweet, A., Venkataraman, A., Stufflebeam, S.M., Liu, H., Tanaka, N., Golland, P., 2013. Detecting Epileptic Regions Based on Global Brain Connectivity Patterns. In: *MICCAI: Medical Image Computing and Computer Assisted Intervention*. Nagoya, Japan, LNCS, pp. 98–105.

Tononi, G., Sporns, O., Edelman, G.M., 1994. A measure for brain complexity: relating functional segregation and integration in the nervous system. PNAS 91, 5033–5037.

Tzourio-Mazoyer, N., Landeau, B., Papathanassiou, D., Crivello, F., Etard, O., Delcroix, N., Mazoyer, B., Joliot, M., 2002. Automated anatomical labeling of activations in {SPM} using a macroscopic anatomical parcellation of the {MNI MRI} single-subject brain. Neuroimage 15 (1), 273–289.

Van Dijk, K.R.A., Hedden, T., Venkataraman, A., Evans, K.C., Lazar, S.W., Buckner, R.L., 2010. Intrinsic functional connectivity as a tool for human connectomics: theory, properties, and optimization. J. Neurophysiol. 103 (1), 297–321.

Venkataraman, A., Van Dijk, K.R.A., Buckner, R.L., Golland, P., 2009. In: *Exploring Functional Connectivity in fMRI via Clustering. ICASSP: International Conference on Accoustics, Speech and Signal Processing*. IEEE, pp. 441–444.

Venkataraman, A., Kubicki, M., Westin, C.-F., Golland, P., 2010. In: *Robust Feature Selection in Resting-State fMRI Connectivity Based on Population Studies. MMBIA: IEEE Computer Society Workshop on Mathematical Methods in Biomedical Image Analysis*. IEEE, pp. 1–8.

Venkataraman, A., Kubicki, M., Golland, P., 2013a. From brain connectivity models to region labels: identifying foci of a neurological disorder. IEEE Trans. Med. Imaging 32 (11), 2078–2098.

Venkataraman, A., Kubicki, M., Golland, P., 2013b. From brain connectivity models to region labels: Identifying foci of a neurological disorder. IEEE Trans. Med. Imaging 32 (11), 697–704.

Venkataraman, A., Duncan, J.S., Yang, D., Pelphrey, K.A., 2015. An unbiased Bayesian approach to functional connectomics implicates social-communication networks in autism. NeuroImage Clin. 8, 356–366.

Venkataraman, A., Yang, D., Pelphrey, K.A., Duncan, J.S., 2016a. Bayesian community detection in the space of group-level functional differences. IEEE Trans. Med. Imaging 35 (8), 1866–1882.

Venkataraman, A., Yang, D.Y.-J., Dvornek, N., Staib, L.H., Duncan, J.S., Pelphrey, K.A., Ventola, P., 2016b. Pivotal response treatment prompts a functional rewiring of the brain among individuals with autism spectrum disorder. Neuroreport 1–5.

Venkataraman, A., Wymbs, N., Nebel, M., Mostofsky, S., 2017. In: *A Unified Bayesian Approach to Extract Network-Based Functional Differences from a Heterogeneous Patient Cohort. CNI: International Workshop on Connectomics in NeuroImaging*. Springer, pp. 1–8.

Waterhouse, L., Gilberg, C., 2014. Why autism must be taken apart. J. Autism Dev. Disord. 1–5. Retrieved from, https://doi.org/10.1007/s10803-013-2030-5.

Watson, R., Latinus, M., Charest, I., Crabbe, F., Belin, P., 2014. People-selectivity, audiovisual integration and heteromodality in the superior temporal sulcus. Cortex 50, 125–136. https://doi.org/10.1016/j.cortex.2013.07.011.

Wildgruber, D., Ackermann, H., Kreifelts, B., Ethofer, T., 2006. Cerebral processing of linguistic and emotional prosody: fMRI studies. Prog. Brain Res. 156, 249–268. https://doi.org/10.1016/j.neuroimage.2005.09.059.

Williams, J.H.G., Waiter, G.D., Perra, O., Perrett, D.I., Whiten, A., 2005. An fMRI study of joint attention experience. Neuroimage 25 (1), 133–140. https://doi.org/10.1016/j.neuroimage.2004.10.047.

Wolff, J.J., Gu, H., Gerig, G., Elison, J.T., Styner, M., Gouttard, S., Botteron, K.N., Dager, S.R., Dawson, G., Estes, A.M., Evans, A.C., Hazlett, H.C., Kostopoulos, P., RC, M.K., Paterson, S.J., Schultz, R.T., Zwaigenbaum, L., Piven, J., the IBIS Network, 2012. Differences in white matter fiber tract development present from 6 to 24 months in infants with autism. Am. J. Psychiatry 169, 589–600.

Xia, M., Wang, J., He, Y., 2013. BrainNet viewer: a network visualization tool for human brain connectomics. PLoS One 8, e68910. https://doi.org/10.1371/journal.pone.0068910.

Yang, D., Rosenblau, G., Keifer, C., Pelphrey, K.A., 2015. An integrative neural model of social perception, action observation, and theory of mind. Neurosci. Biobehav. Rev. 51, 263–275. Retrieved from, http://view.ncbi.nlm.nih.gov/pubmed/25660957.

Yarkoni, T., Poldrack, R.A., Nichols, T.E., Van Essen, D.C., Wager, T.D., 2011. Large-scale automated synthesis of human functional neuroimaging data. Nat. Methods 8, 665–670. https://doi.org/10.1038/nmeth.1635.

Insights into cognition from network science analyses of human brain functional connectivity: Working memory as a test case

2

Dale Dagenbach

Department of Psychology, Wake Forest University, Winston-Salem, NC, United States

CHAPTER OUTLINE

INTRODUCTION

The previous decade has seen existing and emerging tools of network science studies of human brain functional connectivity deployed in efforts to characterize the relationship between the brain and various aspects of cognition. Those studies have yielded a wealth of new data for cognitive neuroscientists to ponder over, and a new wave of even more complex data is emerging as advances in analyses focused on capturing dynamic components of human brain connectivity start to bear fruit. Given all of that new information, it seems appropriate to try to evaluate what we have learned thus far. To that end, this chapter reviews insights regarding cognition that graph-theory-based network science analyses of human brain functional connectivity studies using functional magnetic resonance imaging (fMRI) data have yielded to date and some promissory notes they might deliver on in the near future. This review focuses on the extent of studies of working memory for reasons noted later and notes possible areas where further investigation might be fruitful.

Connectomics. https://doi.org/10.1016/B978-0-12-813838-0.00002-9

WORKING MEMORY

The choice of working memory as a test case for evaluating findings from studies of human brain functional connectivity was predicated on multiple factors. First and foremost, working memory has been a central construct in most theories of human cognition since Baddeley and Hitch's (1974) seminal characterization. Baddeley and Hitch developed the idea of working memory to highlight the importance of the transient storage and active manipulation of information in cognition, proposing a system in which these interacted. In their original model, working memory was characterized as a mental workspace consisting of a "central executive" used for manipulation of information, along with separate buffer systems for the temporary storage of visuospatial and verbal information. The "phonological loop" and "visuospatial sketchpad" buffers were posited to be separate storage systems based on a relative lack of dual task interference for verbal versus visuospatial processing in behavioral experiments, although both were linked to the same central executive.

Other more recent accounts of working memory continue to emphasize the idea that working memory allows us to transiently hold a limited amount of information in a manner that allows for access of it in the absence of an external stimulus's presence, although exact details of the system may vary considerably. In contrast to Baddeley's model with multiple components of working memory, another prominent class of models is exemplified by Cowan's (1988) conceptualization of working memory as the central focus of attention that can be directed either toward sensory or long-term memory representations. A similar emerging theme across more recent studies would be a characterization of working memory as a limited capacity system intrinsically linked to attention but with great variability in terms of its contents and where the representations for those contents are stored in the brain (D'Esposito and Postle, 2015). And it remains the case that, for many current accounts, working memory is still characterized as "the workbench of consciousness" (Marois, 2015).

Befitting its theoretical importance, working memory is among the most highly studied constructs in human cognition; extensive research programs have examined topics such as verbal working memory, visual working memory, and spatial versus object working memory with literally tens of thousands of relevant studies. However, despite this extensive research, a core aspect of working memory remains only partly understood: the concept that working memory is in some way a limited capacity system in which only a relatively small amount of information can be held in a highly accessible state is common across nearly all accounts. That concept seems to capture an essential limitation in our cognitive capabilities, but the mechanism(s) underlying that limitation remain to be fully elucidated. Candidates for the cause of capacity limitations include the decay of working memory representations over time, some form of attentional resource shared with other processing operations, and interference between working memory representations that are not assumed to decay, among others (Oberauer et al., 2016; D'Esposito and Postle, 2015). Given the incomplete resolution of this issue using behavioral studies and activation-based cognitive neuroscience methods, one interesting question for the present review is the extent to which network science analyses might provide any further insight about the source of working memory capacity limits.

This is not an easy issue to resolve. For one thing, the measurement of working memory capacity is complex. Some forms of assessment measure primary memory or the amount of information that can be held in immediate attentional focus (Shipstead et al., 2014.) The capacity of working memory measured in this way is estimated to be approximately three to five units of information for adults for verbal working memory if strategies such as chunking and rehearsal are not used (Cowan et al., 2012). Similarly, for visual working memory, capacity estimates of four objects are common, although that apparent capacity may decline with increased object complexity (Luck and Vogel, 2013). However, other measures of working memory require measurement of not only the information in immediate conscious awareness (primary memory) but that also may reflect the use of a cue-dependent search mechanism in secondary memory (Unsworth and Engle, 2007). Secondary memory contains information relevant for the current context not in primary memory, and the extent to which the search is focused or constrained affects working memory performance.

Although it is a difficult issue to resolve, understanding the source of capacity limits would seem central to any full theory of cognition. The importance of working memory capacity is highlighted in research on individual differences in cognition. Individual differences in working memory capacity have been linked to multiple cognitive measures including differences in fluid intelligence (Kane et al., 2004), language comprehension (Daneman and Mirikle, 1996), performance on academic achievement tests (Cowan et al., 2005), and real world tasks such as problem solving (Engle and Kane, 2004), among others. In addition to predicting individual differences in various aspects of cognitive functioning and working memory, working memory capacity is also often invoked as an important determinant of developmental changes in functioning and has been used to explain, in part, intellectual development during childhood (Bayliss et al., 2003) and intellectual decline during aging (Park et al., 2002).

Beyond its central role in cognitive theory and the intriguing question of where capacity limits arise from, another reason for using working memory as a test case for looking at how network science analyses of human brain functional connectivity can enhance our understanding of cognition is that there already is an extensive cognitive neuroscience literature on working memory. The vast majority of these studies have not employed network science methods but rather have examined areas of brain activation with the goal of localization of function of various components of working memory. This leads to the question of whether network science human brain functional connectivity analyses expand our understanding of working memory in other important ways in addition to any light that they might shed on the characterization of capacity limitations. In a recent review of the possible applications and utility of network science analyses for cognitive neuroscience, Medaglia et al. (2015) suggest that one means for evaluating novel approaches in science is by determining the extent to which they lead to the discovery of "novel processes, structures, and phenomena that assist us in interpreting … prior empirical or principled knowledge." One way that network science analyses might expand upon this existing knowledge would be by either allowing identification of other as yet unidentified brain regions that contribute to working memory or enhancing understanding of the roles of already identified regions.

The existing cognitive neuroscience literature typically characterizes working memory as a network of distributed brain areas, although with considerable variability in exactly which areas. In a recent review of the neural instantiation of working memory, Eriksson et al. (2015) adopted a process-oriented view of working memory in which prefrontal cortex (PFC) and parietal cortex activity is tied to executive and manipulation processes. These executive and manipulation processes can be linked to content-related representations almost anywhere in the brain, including sensory or long-term memory representations. Additionally, the organization of executive demands within the PFC seems somewhat variable, although there is a suggestion of a rostral to ventral organization in dorsal PFC with increasing abstraction of tasks and goals as one moves to more rostral areas. Then parietal cortex also is associated with executive processes in working memory and capacity, with activity increasing as the number of items maintained increases up to its peak of three or four items. Eriksson et al. also note possible roles for the cerebellum and basal ganglia. This existing widespread network characterization of working memory would seem well-suited for exploration using network science functional connectivity analyses, which in turn could provide further insights by showing how interactions between these already identified regions and other as yet unidentified brain regions are critical to performance. In other words, functional connectivity between the areas previously identified through activation-focused research and areas not found in that research could mediate working memory performance (see Stanley et al., this volume, for a review of this argument applied toward episodic memory). In the case of working memory, such findings might help elucidate the roles of areas such as the cerebellum and basal ganglia.

TASKS FOR STUDYING WORKING MEMORY

Consideration of the cognitive neuroscience of working memory literature also brings into focus the importance of the tasks used to study working memory. Much of the existing cognitive neuroscience literature has employed the n-back task in which stimuli, typically letters, are presented serially, and the participant must indicate whether the current item is the same as the n-th back item, with n's ranging from 0 to 3. Because it allows for the observation of the same system at different levels of demand, the n-back paradigm has been viewed as an ideal vehicle for studying working memory in studies using fMRI.

However, working memory is a complex construct, and behavioral research suggests that using other versions of working memory tasks might lead to additional important insights. In addition to n-back tasks, other measures of working memory include complex span tasks, in which two tasks are typically interleaved, and running span tasks. An example of a complex span task would be an operation span in which a simple math problem and result is presented for verification ($2 \times 9 = 16$), followed by a word to be remembered. The running span task presents a series of letters or words serially, with periodic requests to recall the last three to seven items. These additional tasks are worth considering because much of the literature on individual differences that links working memory capacity to other intellectual functions has used complex span tasks. Further, research suggests that the correlation between *n*-back and complex span is relatively weak, although both do correlate strongly with fluid intelligence. The correlation between running span tasks and fluid intelligence is even stronger (Shipstead et al., 2016). Nearly all studies of working memory using task-based human brain functional connectivity analyses have relied on simpler measures such as the *n*-back task that may not capture important aspects of working memory. The few activation-focused neuroimaging studies that have used complex span tasks have implicated medial temporal neural regions not typically seen in studies using the *n*-back procedure (Chein et al., 2011; Faracco et al., 2011). The reliance on the *n*-back procedure in most studies may have reflected the difficulty of implementing more complex span tasks in the neuroimaging environment, but there are variations of complex span that would seem amenable to testing in the scanner that have shown good psychometric properties (Redick et al., 2012).

INSIGHTS FROM NETWORK SCIENCE ANALYSES OF HUMAN BRAIN FUNCTIONAL CONNECTIVITY RESTING STATE DATA

For approximately a decade, human brain functional connectivity research was dominated by the use of intrinsic or resting state data. As elaborated on throughout this volume, that approach yielded numerous insights. More recently, however, studies of functional connectivity while participants perform various tasks have begun to proliferate. The resulting contrasts of resting state networks and task-state networks have yielded both remarkable similarities and intriguing differences. Among the interesting similarities are suggestions that resting state functional connectivity data can be used to predict individual differences in task-state brain activity (Tavor et al., 2016; Cole et al., 2016). Among the intriguing differences are findings that the locations of hub nodes (nodes with a high degree or a high number of connections to other nodes) shifts between resting state and task states. In one recent large sample review, resting state hub nodes were concentrated in primary sensory and motor regions, whereas during a performance of a variety of tasks, that concentration shifted to salience, fronto-parietal, ventral attention, and cingulo-opercular control modules (Bolt et al., 2017).

For studies of cognition, the decision to focus on functional connectivity data obtained during resting state or a task-state creates an interesting quandary. It makes sense to take the application of resting state data to the understanding of cognition as far as possible, if only because of the pragmatic considerations. If the brain reconfigures in important ways for each cognitive process or activity, one can imagine functional neuroimaging studies looking at the consequences of hundreds, if not thousands, of manipulations of working memory paradigms, let alone for other areas of cognition such as language processing, long-term memory, decision making, etc.

The behavioral science literature and, to some extent, the activation-focused cognitive neuroscience literature suggest that relatively minor tweaks to experimental procedures within the same cognitive domain can have major effects on the results. The sheer effort and cost of such studies would be staggering. If important significant insights into various aspects of cognition can be derived from resting state data, then the same set of functional neuroimaging data collected once, in conjunction with sage behavioral data collection, could yield advances in understanding across many different areas of cognition. On the other hand, it is clear that significant and important changes in brain functional connectivity do occur in task states compared with resting state, therefore it seems likely that understanding these will be a critical part of our understanding of the relationship between functional connectivity and cognition. And there is also the intriguing possibility that the relationships between resting state and task-state functional connectivity brain networks may provide important insights: how much the brain is able to, or has to, reconfigure from rest to task depends in part upon where it starts from. Ultimately, it seems likely that a conjunction of human brain functional connectivity analyses using both resting-state and task-state data will yield the most answers (see Campbell and Schacter, 2017, and associated commentaries for a recent discussion in the context of studying cognitive aging). Therefore, network science functional connectivity data analyses looking at both resting-state and task state data are reviewed later.

RESTING STATE FUNCTIONAL CONNECTIVITY AND WORKING MEMORY

Modularity

A number of studies of working memory and human brain functional connectivity have identified modularity as a property of the brain that may be systematically related to working memory performance. This is true of both modular organization of the brain in resting state as well as when performing a working memory task. Modules in brain networks are clusters of densely connected nodes or network communities (Sporns and Betzel, 2016). Generally, an overall network is subdivided into modules consisting of nodes strongly interconnected within a module but only weakly connected to nodes in other modules. There are a number of different approaches for doing this that make different assumptions, but generally the goal is to achieve some modularity maximization such as attaining the highest possible value of the modularity quality function, (Newman and Girvan, 2004). Because absolute modularity maximization is computationally infeasible (NP-hard), it remains a challenge to find the right procedures for achieving the best possible fit. Theoretical issues also exist regarding whether nodes can belong to multiple modules: how "good" a modular partition should be for the network to be viewed as having a modular structure, how to select the best modular map given that multiple analyses of the same data will yield different solutions, etc. (see Sporns and Betzel, 2016, for an extended and accessible discussion of relevant issues).

Despite many unresolved issues concerning modularity analyses, the existing literature does suggest it is a potentially fruitful area for exploration in both resting-state and task-state networks. In one particularly intriguing study, Stevens et al. (2012) assessed the relationship between resting state modularity and visual short-term memory (VSTM) capacity using functional brain networks comprised of 34 regions previously associated with a variety of tasks. Participants performed a VSMT task to assess capacity, followed by a resting-state fMRI scan, on two separate occasions. Modularity was significantly positively correlated with VSTM capacity at both sessions. At the same time, modularity also was sensitive to within-individual changes over time; changes in modularity were significantly correlated with changes in VSTM across the two sessions.

Other studies have since found more variable results. Alavesh et al. (2015) addressed the relationship between resting-state modularity and working memory performance using a computational span measure characterized as a measure of numerical working memory, and a visuospatial working memory measure. Resting-state modularity (Q) correlated significantly with visuospatial working memory performance but not with the computational span measure. A complex analysis by Yue et al. (2017) correlated modularity from resting-state functional connectivity with behavioral performance on a simple span task (digit span) and a complex span task (operation span), along with a number of other cognitive measures. Behavioral data were characterized as being from either complex tasks or simple tasks, with both digit span and operational span assigned to the complex task category. Correlations between modularity and the composite scores for the simple and complex tasks revealed a significant positive correlation between modularity and simple task performance but a marginally negative one between modularity and complex task performance. Moreover, looking at the correlations between modularity and the individual tasks, there was a significant negative correlation between modularity and operation span, and a nonsignificant negative correlation between modularity and digit span.

The variety of results obtained from looking at resting-state modularity and working memory capacity suggests that the story is likely to be complex, and results may hinge on many as yet undetermined methodological questions, including the methods used to construct networks, the methods used to calculate modularity, and the exact nature of the working memory task. Interestingly, the findings of a positive relationship between resting-state modularity and working memory performance both came when measures of visual or visuospatial working memory performance were used. Further resolution of this relationship would be a valuable contribution. One possibility, albeit unsupported by empirical evidence at this point, is that visual processing during eyes-open resting-state data collection imposes some organization on modularity reflected in the resulting community structure.

FUNCTIONAL CONNECTIVITY DURING WORKING MEMORY TASK PERFORMANCE

Although the resting-state data suggest some possible links between network science analyses of functional connectivity in the brain and working memory performance, examination of changes in brain organization as people perform working memory tasks provides additional information. First, a number of studies have shown that organization does change in meaningful ways between rest and task-state networks. For example, Rzucidlo et al. (2013) compared whole brain network organization using voxel-based brain networks and a repeated measures design in which participants repeatedly transitioned between resting state and a 2-back working memory task, allowing for examination of both stability and changes within and between rest and working memory states. Whole brain analyses indicated a significant decrease in local efficiency between rest and working memory, and regional analyses indicated a significant decrease in local efficiency and degree for the same comparison in the precuneus, typically thought of as part of the default mode network. Conversely, degree increased significantly in the dorsolateral prefrontal cortex (DLPFC), a region generally associated with working memory. Stanley et al. (2014) were able to relate changes in whole brain global and local efficiency in brain networks to working memory performance in a study that included younger and older adults. Whole brain local efficiency decreased significantly between rest and 2-back for the younger participants and marginally for the older participants. Local and global efficiency contributed significantly to a regression model predicting working memory performance using the task-state data but not when resting-state data were used. Increases in global efficiency during task compared with rest were associated with higher working memory performance for young adults but not for older adults.

Modularity in task-based networks also has been linked to working memory task demands. Stanley et al. (2014) explored the modular and modular hub organization of a sample of young adults during performance of both a 1-back and 2-back task, thus allowing examination of working memory load effects. Whole brain modularity (Q) did not differ significantly as a function of working memory load, but functional cartography analyses (Guimera and Nunes Amaral, 2005) identifying provincial and connector hubs within default mode network and working memory network regions indicated significant increases in connector hubs for both regions between 1- and 2-back. Provincial hubs are high-degree nodes with mostly intramodular edges or connections, whereas connector hubs are high-degree nodes with a substantial quantity of intermodular connections. Further examination of the relationship between changes in modularity and changes in working memory performance between 1-back and 2-back suggested that decreased modularity with increased working memory load was associated with improved performance. Additionally, although working memory performance declined in the harder 2-back task, decreases in provincial hubs and increases in connector hubs between 1-back and 2-back were associated with less of a decrement.

The same pattern of increased working memory load being associated with less modular, more integrated organization was also found in a study contrasting rest, sequence tapping, and n-back (Cohen and D'Esposito, 2016). In this case, modularity during rest and sequence tapping were not statistically different, but modularity during the n-back task was significantly lower. The balance within and between module connections also was assessed using a measure of system segregation that indicated higher segregation in the sequence tapping task compared with the n-back task.

Whole brain local efficiency decreased in both tasks compared with rest, whereas global efficiency was significantly higher in n-back compared with the sequence tapping task, which was numerically but not statistically significantly lower compared with rest. Explorations of connector and provincial hubs found a significant increase in the number of connector hubs during the n-back compared with rest and sequence tapping, whereas the number of connector hubs decreased but not significantly from rest to sequence tapping. In terms of relations to behavioral performance, modularity (Q) was negatively correlated with accuracy on the 3-back task, as was local efficiency. Global efficiency was positively correlated, as was the number of connector hubs. These findings, along with those of Stanley et al. (2014), are consistent with the view that less modular or greater cross-network organization is related to working memory performance.

Another issue worth considering is the nature of modularity change as the brain transitions from resting state to a task state. One preliminary study looking at this issue used the data from the Rzucidlo et al. (2013) repeated measures design that alternated between rest and a 2-back task (Dagenbach et al., 2014). This design allowed asking a question about whether modular organization would change as one went from rest to n-back, perhaps a given, but also the consistency of that change. An initial question was how to define whether a module had changed. We chose to do this at a nodal level with the arbitrary definition that a node's module had changed if more than 50% of its neighbors had changed from time 1 to time 2. Note that a node's module could change using this definition based on other nodes leaving the module, on new nodes joining the module, or on a combination of these factors. Subsequently, more elegant algorithms for calculating modularity change have been developed, but the choice of this simple one provided a surprising outcome. The initial analysis comparing rest with n-back was somewhat surprising, suggesting that modularity, at least when created using voxel-based networks and examined using our nodal level metric, was not stable at all. Virtually all nodes had changed modules using that arbitrary definition. Fortunately, the repeated measures design allowed for the obvious next comparison—rest at time 1 to rest at time 2, etc., and working memory at time 1 to working memory at time 2, etc. The results were nearly as variable but with a critical difference: there was evidence of consistency in modular membership across sessions and participants for regions associated with working memory in the n-back task data, and evidence of consistency in visual and motor areas in the resting-state data, along with a hint of consistency for the precuneus, an area strongly associated with the default mode network (see Figs. 1 and 2). Thus, looking at the consistency of change in modularity seemed to provide a roundabout way of seeing either localization of function or something related.

The studies reviewed thus far used functional connectivity networks in which each edge or connection between nodes had a single value. Looking at dynamic functional connectivity, or how networks change over time, is an emerging research strategy. One illustration of this applied to working memory comes from an elegant study that contrasted dynamic functional connectivity during the performance of a task with significant working memory demands (2-back) to a control

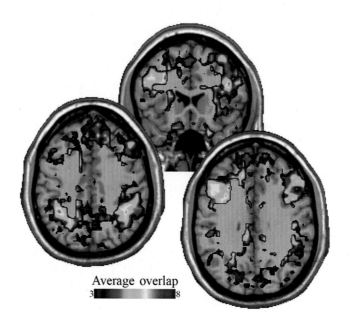

FIG. 1

Nodes showing consistent modular membership across participants and sessions during a working memory task.

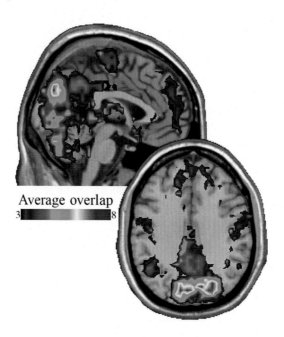

FIG. 2

Nodes showing consistent modular membership across participants and sessions during resting state.

condition with fewer demands (0-back) (Braun et al., 2015). Using an overlapping sliding time window approach, they looked at changes in modularity over time as participants alternated between the 0-back and 2-back conditions every 30 s. Multilayered networks were constructed based on data from consecutive time windows, then modules were identified using a dynamic community connection procedure. A metric of flexibility was developed based on the probability that any given brain region changed its membership in functional modules between successive time windows. Flexibility was at its peak for time periods when participants were primarily in either the 0-back or 2-back task but also varied between the two tasks by brain region; frontal regions showed greater flexibility during performance of the 2-back task. A measure of integration based on the number of links between nodes in two different modules showed higher integration in frontal systems for the 2-back task compared with the 0-back task, wherein integration was similar for frontal and occipitoparietal systems. Frontal system flexibility and integration were correlated with accuracy on the 2-back task.

These results also seem in contradiction to the preliminary study reported earlier (Dagenbach et al., 2014). Both studies suggest that modularity change may be related to working memory in interesting ways, but Braun et al. found that high flexibility is related to modular membership during the 2-back task, whereas Dagenbach et al. observed consistent modular membership only for nodes in areas possibly associated with working memory. These contrasting results again may stem from different methods used to construct the networks: the number of nodes, the way in which modularity was calculated, the use of dynamic networks over short periods of time versus contrasting networks constructed over longer periods of time at different time points, etc. How all of these decisions change our characterization of the relationship of modularity change to working memory remains an open question.

CONCLUSIONS

The goal of this chapter was to review the contributions that network science analyses of human brain functional connectivity have made thus far to our understanding of human working memory and to consider possible future directions. From the literature reviewed, it seems clear that these analyses have clearly advanced our understanding of working memory in at least one way and have provided interesting clues in several others. In the absence of the functional connectivity studies, it was already clear from the extensive literature that working memory involved a network of brain regions. What the functional connectivity studies have added to this story is detail about that process: the transition from a more modular organization at rest to a more integrated organization during performance of a working memory task and the extent to which that transition is associated with enhanced performance on the working memory task marks an advance in our understanding. That finding, across a number of studies and specific methodologies, contrasts with a couple of findings linking greater modularity at rest with better VSTM or visuospatial working memory performance (Stevens et al., 2012; Alavesh et al., 2015). One possible resolution might be that greater modularity at rest allows for greater transition to a less modular, more integrated state, without going so far in that direction as to arrive at a state in which performance is actually impaired. In other words, a state in which the brain can alternate more strongly between highly modular organization and highly integrated organization depending on task demands might be optimal. However, that reason cannot explain the inverse relationship between resting-state modularity and operation span performance found in Yue et al. (2017). Given the scarcity of data on the question of whether modular organization at rest predicts working memory performance, and the possible contradictions in those data that exist, this would seem to be another area ripe for further exploration. This exploration might easily be done using some of the archival data sets now in existence such as the Human Connectome Project (Van Essen et al., 2013). Resolving this might provide insights into the fundamental question of understanding the capacity limits associated with working memory.

On a negative note, the network science research reviewed failed to provide clear evidence of the involvement of additional brain regions in working memory. It also failed to provide further insights into the functions of some areas noted in activation-based research such as the basal ganglia and cerebellum. These issues remain open to further exploration.

Another point worth noting is the extent to which the task-based functional connectivity studies have relied on the *n*-back paradigm for studying working memory. This is understandable, but given the behavior studies differences between *n*-back and complex span tasks, further exploration of functional connectivity during the performance of complex span tasks is clearly warranted.

In addition to suggesting that the utility of new scientific approaches might be evaluated by the extent to which they bring new understanding of established data or phenomena, Medaglia et al.'s (2015) review also suggests that they may be evaluated by the extent to which they lead to the discovery of fundamental organizing principles and reveal validated relationships with other variables. The shift from more modular to more integrated states as the brain transitions between resting state or simple tasks and more complex tasks might fulfill the first of those, and numerous studies have used network science functional connectivity analyses to predict a number of disease states in fulfillment of the second. Yet another criterion for evaluating the utility of a new approach in science may be the extent to which it is able to generate new and ever more interesting questions as we learn more. The application of network science functional connectivity analyses to understanding working memory thus far has generated a host of unanswered questions that may need to be resolved using analyses across multiple spatial and temporal scales, and using further dynamic network analyses and methods. It would seem that we have just begun to tap the potential of network science functional connectivity models to inform us about working memory and, by implication, other aspects of cognition as well.

REFERENCES

Alavesh, M., Doebler, P., Holling, H., Thiel, C., Giessing, C., 2015. Is functional integration of resting state brain networks an unspecific biomarker for working memory performance? Neuroimage 108, 182–193.

Baddeley, A.D., Hitch, G., 1974. Working memory. Psychol. Learn. Motiv. 8, 47–89.

Bayliss, D.M., Jarrold, C., Gunn, D.M., Baddeley, A.D., 2003. The complexities of complex span: explaining individual differences in working memory in children and adults. J. Exp. Psychol. Gen. 132 (1), 71–92.

Bolt, T., Nomi, J.S., Rubinov, M., Uddin, L.Q., 2017. Correspondence between evoked and intrinsic functional brain network configuration. Hum. Brain Mapp. 38 (4), 1992–2007.

Braun, U., Schäfer, A., Walter, H., Erk, S., Romanczuk-Seiferth, N., Haddad, L., Schweiger, J.I., Grimm, O., Heinzt, A., Tost, H., Meyer-Lindenberg, A., Bassett, D.S., 2015. Dynamic reconfiguration of frontal brain networks during executive cognition in humans. PNAS 112 (37), 11678–11683.

Campbell, K.L., Schacter, D.L., 2017. Ageing and the resting state: is cognition obsolete? Lang. Cogn. Neurosci. 32 (6), 661–668.

Chein, J.M., Moore, A.B., Conway, R.A., 2011. Domain-general mechanisms of complex working memory span. Neuroimage 54 (1), 550–559.

Cohen, J.R., D'Esposito, M., 2016. The segregation and integration of distinct brain networks and their relationship to cognition. J. Neurosci. 36 (48), 12083–12094.

Cole, M.W., Ito, T., Bassett, D.S., Schultz, D.H., 2016. Activity flow over resting-state networks shapes cognitive task activations. Nat. Neurosci. 19 (12), 1717–1726.

Cowan, N., 1988. Evolving conceptions of memory storage, selective attention, and their mutual constraints within the human information-processing system. Psychol. Bull. 104 (2), 163–191.

Cowan, N., Elliott, E.M., Saults, J.S., Morey, C.C., Mattox, S., Hismjatullina, A., Conway, R.A., 2005. On the capacity of attention: Its estimation and its role in working memory and cognitive aptitudes. Cogn. Psychol. 51 (1), 42–100.

Cowan, N., Rouder, J.N., Blume, C.L., Saults, J.S., 2012. Models of verbal working memory capacity: what does it take to make them work? Psychol. Rev. 119 (3), 480–499.

D'Esposito, M., Postle, B.R., 2015. The cognitive neuroscience of working memory. Annu. Rev. Psychol. 66, 115–142.

Dagenbach, D., Rzucidlo, J.K., Laurienti, P.J., Lyday, R.G., 2014. In: Localization of function with fMRI data but without subtraction: using network science analyses to discern task-related brain networks. Presented at Annual Meeting of the Psychonomic Society, Long Beach, CA.

Daneman, M., Merikle, P.M., 1996. Working memory and language comprehension: a meta-analysis. Psychon. Bull. Rev. 3 (4), 422–433.

Engle, R.W., Kane, M.J., 2004. Executive attention, working memory capacity, and a two-factor theory of cognitive control. In: Boss, B. (Ed.), The Psychology of Learning and Motivation. vol. 44. pp. 145–199.

Eriksson, J., Vogel, E.K., Lansner, A., Bergstrom, F., Nyberg, L., 2015. The neurocognitive architecture of working memory. Neuron 88 (1), 33–46.

Faracco, C.C., Unsworth, N.L., Langley, J., Terry, D., Li, K., Zhang, D., Liu, T., Miller, L.S., 2011. Complex span tasks and hippocampal recruitment during working memory. Neuroimage 55 (2), 773–787.

Guimera, R., Nunes Amaral, L.A., 2005. Functional cartography of complex metabolic networks. Nature 433 (7028), 895–900.

Kane, M.J., Hambrick, D.Z., Tuholski, S.W., Wilhelm, O., Payne, T.W., Engle, R.W., 2004. The generality of working memory capacity: a latent-variable approach to verbal and visuospatial memory span and reasoning. J. Exp. Psychol. Gen. 133 (2), 189–217.

Luck, S.J., Vogel, E.K., 2013. Visual working memory capacity: from psychophysics and neurobiology to individual differences. Trends Cogn. Sci. 17 (8), 391–400.

Marois, R., 2015. The brain mechanisms of working memory: an evolving story. In: Lefebvre, C., Martinez-Trujillo, J., Jolicoeur, P. (Eds.), Attention and Performance XXV. pp. 23–31.

Medaglia, J.D., Lynall, M.E., Bassett, D.S., 2015. Cognitive network neuroscience. J. Cogn. Neurosci. 27 (8), 1471–1491.

Newman, M.E.J., Girvan, M., 2004. Finding and evaluating community structure in networks. Phys. Rev. E 69, 026113.

Oberauer, K., Farrell, S., Jarrold, C., Lewandowsky, S., 2016. What limits working memory capacity? Psychol. Bull. 142 (7), 758–799.

Park, D.C., Lautenschlager, G., Hedden, T., Davidson, N.S., Smith, A.D., Smith, P.K., 2002. Models of visuospatial and verbal memory across the adult life span. Psychol. Aging 17, 299–320.

Redick, T.S., Broadway, J.S., Meier, M.E., Kuriakose, P.S., Unsworth, N., Kane, M.J., Engle, R.W., 2012. Measuring working memory capacity with automated complex span tasks. Eur. J. Psychol. Assess. 28, 164–171.

Rzucidlo, J.K., Roseman, P.L., Laurienti, P.J., Dagenbach, D., 2013. Stability of whole brain and regional network topology within and between resting and cognitive states. PLoS One 8 (8), e70275. https://doi.org/10.1371/journal.pone.0070275.

Shipstead, Z., Lindsey, D.R.B., Marshall, R.L., Engle, R.W., 2014. The mechanisms of working memory capacity: primary memory, secondary memory, and attention control. J. Mem. Lang. 72, 116–141.

Shipstead, Z., Harrison, T.L., Engle, R.W., 2016. Working memory capacity and fluid intelligence: maintenance and disengagement. Perspect. Psychol. Sci. 11 (6), 771–799.

Sporns, O., Betzel, R.F., 2016. Modular brain networks. Annu. Rev. Psychol. 67, 613–640.

Stanley, M.L., Dagenbach, D., Lyday, R.G., Burdette, J.H., Laurienti, P.J., 2014. Changes in global and regional modularity associated with increasing working memory load. Front. Hum. Neurosci. https://doi.org/10.3389/fnhum.2014.00954.

Stevens, A.A., Tappon, S.C., Garg, A., Fair, D.A., 2012. Functional brain network modularity captures inter- and intra-individual variation in working memory capacity. PLoS One 7 (1), e30468. https://doi.org/10.1371/journal.pone.0030468.

Tavor, I., Parker Jones, O., Mars, R.B., Smith, S.M., Behrens, T.E., Jbabadi, S., 2016. Task-free MRI predicts individual differences in brain activity during task performance. Science 352 (6282), 216–220.

Unsworth, N., Engle, R.W., 2007. The nature of individual differences in working memory capacity: active maintenance in primary memory and controlled search in from secondary memory. Psychol. Bull. 114, 104–132.

Van Essen, D.C., Smith, S.M., Barch, D.M., Behrens, T.E.J., Yacoub, E., Ugurbil, K., for the WU-Minn HCP Consortium, 2013. The WU-minn human connectome project: an overview. Neuroimage 80, 62–79.

Yue, Q., Martin, R.C., Fischer-Baum, S., Ramos-Nunez, A.I., 2017. Brain modularity mediates the relation between task complexity and performance. J. Cogn. Neurosci. 29 (9), 1532–1546.

FURTHER READING

Hicks, K.L., Foster, J.L., Engle, R.W., 2016. Measuring working memory capacity on the web with the online working memory lab (the OWL). J. Appl. Res. Mem. Cogn. 5 (4), 478–489.

Stanley, M.L., Simpson, S.L., Dagenbach, D., Lyday, R.G., Burdette, J.H., 2015. Changes in brain network efficiency and working memory performance in aging. PLoS One 10 (4), e0123950. https://doi.org/10.1371/journal.pone.0123950.

Overlapping and dynamic networks of the emotional brain

3

Luiz Pessoa

Department of Psychology and Maryland Neuroimaging Center, University of Maryland, College Park, MD, United States

CHAPTER OUTLINE

Where are functions instantiated in the brain? From phrenology to the present day, the central goal of neuroscience has been to understand the mapping between structure and function. By and large, the pendulum swings back and forth between "localizationist" and "distributed" visions of how the brain supports mental processes. The disputes of Santiago Ramon y Cajal and Camillo Golgi, foundational to neuroscience as a discipline, provide a good example. Golgi was a staunch holist with regard to brain function, believing that his studies showed that axons were densely intertwined, forming a "reticulum" (a continuous network of nerve cells). Cajal traced the paths of axons and mapped the structure of neuronal cell bodies in detail. Rejecting the idea that axons or dendrites formed a physically linked network, he believed in individual nervous "elements" that were absolutely autonomous (Finger, 1994). Many other fascinating disputes in neuroscience have revolved around the question of localization and the search for the "right scale" or "right units."

Currently, a large focus of research aims to understand how distributed networks of brain regions support complex behaviors (Sporns, 2010; Fornito et al., 2016). The goal of the present chapter is to discuss brain networks important for emotion and motivation (Pessoa, 2013) while focusing on two conceptual issues that inform the study of networks in general: network overlap and dynamics.

BRAIN NETWORKS ARE OVERLAPPING

The typical way to conceptualize a brain network is in terms of a set of unique, *non-overlapping* brain regions. For example, Menon, Uddin, and colleagues suggest that a "salience" network including the anterior insula and the anterior cingulate cortex is involved in attention to external and internal events (Bressler and Menon, 2010; Menon and Uddin, 2010). In a well-cited study, Yeo et al. (2011) investigated cortical networks by partitioning the cortex into 7, 10, 12, or 17 networks. The 7-network estimate, in particular, corresponded well with networks commonly studied in the literature, including visual, somatomotor, dorsal attention, ventral attention, limbic, frontoparietal, and "default" networks. The general approach thus assumes that a brain region belongs to one and only one network. For example, a region in the dorsolateral prefrontal cortex (PFC) that belongs to the frontoparietal network cannot, by definition, participate in the dorsal attention network. ("Networks" or "communities" are used interchangeably here and refer to subsets of brain regions, or possibly to the entire "brain network".)

However, the importance of understanding and characterizing *overlapping structure* has been discussed for some time across disciplines, including sociology (Wasserman and Faust, 1994) and biology (Gavin et al., 2002). For example, biologists exploring protein interactions have found that a substantial fraction of proteins interact with several communities at the same time (Gavin et al., 2002). These and other studies have led to the suggestion that "actual networks are made of interwoven sets of overlapping communities" (Palla et al., 2005, p. 814). Indeed, the study of methods to probe overlapping network structure has been a robust research area, with important developments in the past decade (e.g., Evans and Lambiotte, 2009; Gopalan and Blei, 2013; Xie et al., 2013).

Why, in brain research, do we encounter such emphasis on nonoverlapping networks? One possibility is a lingering overreliance on regions as units that compute specific functions. In this view, brain areas compute a specific function, one that is perhaps elementary and needs other regions to be "actualized", but nonetheless is well defined. Thus, brain regions have a single function, that is, they are not multifunctional (for further discussion, see Pessoa, 2014).

Some proposals are starting to depart from this traditional conceptualization of brain networks as essentially independent sets of brain regions. In their model of cognitive control, Cocchi et al. (2013) discuss the need to extend existing frameworks to account for context-specific, task-dependent interactions among hubs (that is, particularly well-connected regions). They propose that cognitive control is achieved by the flexible emergence of patterns of interaction and competition among regions. In the present chapter, it is proposed that one reason regions may belong to (or affiliate with) multiple networks is that they participate in different "region assemblies" depending on task demands. In this regard, Cole et al. (2013) described how some brain regions flexibly shift their functional connectivity patterns with multiple brain networks across a wide variety of tasks. They suggest that frontoparietal regions are particularly important "flexible hubs."

The framework advanced here proposes that networks contain overlapping regions, such that brain areas will belong to several intersecting networks (Mesulam, 1990). In this manner, the processes carried out by an area will depend on its network affiliation at a given time. What determines a region's affiliation? The importance of the *context* within which a brain region is operating must be considered (McIntosh, 2000). For example, region R will be part of network N_1 in a certain context but will be part of network N_2 in another (Fig. 1A). The existence of context-dependent, overlapping networks also means that, from the perspective of structure-function mappings, a given region will participate in multiple processes.

In a recent study, we investigated the potential overlapping structure of functional brain connectivity in a data-driven fashion (Najafi et al., 2016). Instead of assuming that regions belong to a single network, we allowed them to belong to *all* communities. Thus, when community organization was determined, a node's *participation* to every community was estimated; participation strengths were called *membership values* (Fig. 1B). Each node was assigned a probability-like membership value to each of the existing communities, resulting in a membership vector with entries between 0 and 1 (and summing to 1); entries close to 1 indicated membership to essentially one community, intermediate values indicated membership to multiple communities, and values close to 0 indicated weak or no community affiliation. The approach to detecting overlapping communities we adopted was originally based on the mixed-membership stochastic blocks model (Airoldi et al., 2008) within the context of stochastic variational inference (Hoffman et al., 2013). The specific algorithm

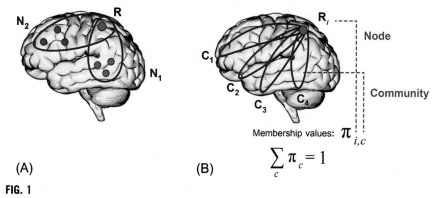

(A)   (B) $$\sum_c \pi_c = 1$$

FIG. 1

Organization of brain networks. (A) The role of context in determining network affiliation: region R will cluster with regions in network N_1 during one context but with regions in network N_2 in another context. (B) In the framework of overlapping networks (or mixed membership), each brain region affiliates with multiple communities with varying strengths called "membership values". For each brain region, these are probability-like values between 0 and 1 that sum to 1 (summed across communities). Here, the membership values of region Ri are based on strengths to each of communities C_1 to C_4.

we employed was developed by Gopalan and Blei (2013), and can be viewed as a generative Bayesian algorithm favoring assortativity (nodes with similar memberships are more likely to cluster together).

When applied to resting-state functional magnetic resonance imaging (fMRI) data, the algorithm captured community organization, matching many of the general features of standard disjoint networks. For comparison purposes, we also determined community structure based on standard, disjoint community assignment (based on k-means clustering). In particular, out of six overlapping communities, four of them exhibited a good degree of similarity with specific disjoint communities; for example, the overlapping community shown in Fig. 2A resembles the default network. Notably, several of the overlapping communities correlated nontrivially with multiple disjoint communities, indicating that mixed communities are not fully captured by a single disjoint community, and that they may contain information that is not well described by disjoint communities.

To gain a richer understanding of the overlapping networks obtained, for each community we considered the entire distribution of membership values, that is, we considered the membership vectors of all brain regions simultaneously. Recall that each brain region was allowed to participate in all communities; however, a high membership value to one community implies that the node can only affiliate weakly with other communities because membership values sum to 1 (Fig. 1B). Thus, a community containing nodes that affiliated mostly with that community would exhibit a membership distribution with values close to 1. Instead, we found that all

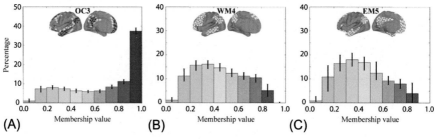

(A)   (B)   (C)

FIG. 2

Overlapping communities during rest and task conditions. Each panel displays the frequency distribution of membership values of all nodes in the brain to one specific community (six overlapping communities were detected at both rest and during tasks). (A) During rest, one of the communities resembled the standard task-negative network (see inset). Although a subset of nodes had membership values close to 1 (indicating that they were more exclusively affiliated with the specific community), a large number of regions had values spanning most of the range of affiliation strengths. (B) Membership values for a representative community obtained during the working memory task. (C) Membership values for a representative community obtained during the emotion task.

Figure based on Najafi, M., Mcmenamin, B.W., Pessoa, L. 2016. Overlapping communities reveal rich structure in large-scale brain networks during rest and task conditions. NeuroImage 135, 92–106.

mixed communities had many regions with nontrivial membership values far from the peak of 1 (Fig. 2A), which indicated that the overlapping communities were not well described by "exclusive nodes" (nodes that participated in only one community) but by "multipartner nodes" (nodes that participated in multiple communities). Even for the most "peaked" community shown in Fig. 2A (and resembling the default network), the bulk of the membership values were below 1 (and most of the probability mass [~60%] was contained in the nonmaximal bins). Thus, whereas the mixed-membership approach reproduced many of the general features of disjoint communities, mixed communities contained information not captured by disjoint communities.

Although network structure has been extensively studied during taskless states, less is known during task execution. In particular, how network organization might be altered by tasks is actively debated (for discussion and references, see Cole et al., 2013). To address this question, we investigated the overlapping network structure of functional MRI data collected during working memory (2-back condition) and emotion (matching emotional faces) tasks (Najafi et al., 2016). In broad terms, several of the communities during the two tasks resembled overlapping communities found at rest. However, examination of the distribution of membership values for individual communities revealed important differences in overlapping information. For the working memory task, five of the six communities exhibited a distribution of membership values that did not peak at higher values of membership (see Fig. 2B), indicating a looser affiliation of nodes to particular communities. In other words, during working memory, most regions affiliated with multiple communities in a(n) (at least) moderate manner. For the emotion task, the distribution of membership values followed a similar pattern (see Fig. 2C), with a more pronounced shift toward intermediate values of membership strength. Taken together, these results indicate that the community organization during tasks differs from that during rest in important ways. Notably, brain regions affiliate with a larger number of networks.

NETWORK MODULARITY

Standard community-detection algorithms are often based on formalizing a network measure called "modularity," based on optimizing a quality function over the possible ways of dividing a network (Newman, 2010). Thus, modularity helps differentiate between "good" and "bad" potential partitions of the nodes of a network. In many cases, the criterion expresses the idea that communities should have nodes that are more heavily interconnected, such that within-community edges are more numerous (or stronger) than between-community edges. Thus, assuming binary networks (with nodes connected or not), if two regions are connected and attributed to the same community, modularity increases, and if two regions are connected but attributed to two different communities, modularity decreases.

Whereas a quality function is frequently defined and optimized to determine a community partition of a network (that is, to help determine good ways of partitioning the nodes), a quality function of interest can also be used to characterize partitions suggested by other methods. This is particularly useful when the partitioning of an overall network was done via a different quality function, or a set of criteria that do not map to a single simple quality function. We proposed a quality function to summarize the community partitions of the overlapping community detection method we employed, and defined the modularity of a community c as follows (Najafi et al., 2016):

$$Q_c = \frac{1}{2m} \sum_c \sum_{i,j \in c} \left(A_{i,j} - \frac{k_i k_j}{2m} \right) \pi_{i,c} \pi_{j,c}$$

where m is the total number of links in the graph, $A_{i,j}$ is the edge between nodes i and j (1 if edge present, 0 otherwise), k_i is the degree of node i, and $\pi_{i,c}$ is the membership value of node i in community c. This extension of modularity is straightforward because the part of the equation prior to the product of membership values ($\pi_{i,c} \pi_{j,c}$) implements standard disjoint modularity (Newman, 2010), and the product of the membership values incorporates their strengths into the standard formulation of modularity. Note that, by definition, standard disjoint modularity considers membership as a binary value, such that the product $\pi_{i,c} \pi_{j,c}$ is either 0 or 1 (nodes are either in the same community or not).

Insight into the changes in network structure linked to tasks states can be gained by studying modularity, as it provides a measure of the extent to which signals potentially can flow between communities. In our study, when compared between rest and task conditions, modularity was highest during resting state and lower for both working memory and emotion tasks, for which modularity scores were fairly similar (Fig. 3). Notably, modularity scores of all individual communities during working memory and emotion were lower than values observed at rest, showing that the reduction was not driven by changes to one or a just a few communities. These findings further support the idea that tasks alter functional connectivity structure in important ways relative to rest. In particular, the decreased modularity is consistent with *increased* communication during task performance.

NODE TAXONOMY: HUBS AND BRIDGES

An important goal of network science is to understand how nodes (here, brain regions) potentially participate in "signal communication" so as to better characterize how network organization supports (or not) particular functions. For example, regions characterized by a high degree of connectivity, often called *hubs*, are believed to be important in regulating the flow and integration of information between regions. Guimera and Nunes Amaral (2005) proposed that hubs can be grouped into distinct subtypes: for example, *provincial* hubs (well-connected nodes with almost all of their links within a single community), *connector* hubs (well-connected nodes with at least half of their links within a community; that is, nodes that participate strongly within a community *and* in communication across communities), and *kinless* hubs (hubs with fewer than half of their links within a community; that is, nodes that may be a conduit for signal propagation across the network but do not participate strongly within a specific community). Classification in terms of hub subtypes is useful for

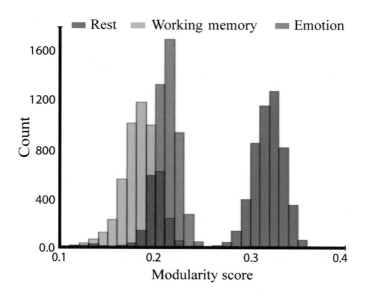

FIG. 3

Modularity scores of overlapping communities during rest and task conditions. The histograms depict the whole-brain modularity (obtained by summing modularity scores across communities). The distributions were obtained via 5000 bootstrapped iterations that resampled over participants.

Reproduced with permission from Najafi, M., Mcmenamin, B.W., Pessoa, L. 2016. Overlapping communities reveal rich structure in large-scale brain networks during rest and task conditions. NeuroImage 135, 92–106.

understanding the organization of networks because each of the defining connectivity patterns is potentially indicative of general functional roles that do not strictly depend on the type of network being studied (social, technological, biological, etc.).

Relative to the case of disjoint networks, the mixed-membership approach provides additional ways to characterize node contributions to network function because each node's description is based on its participation across all networks *simultaneously*. In particular, hub subtypes can be naturally defined by characterizing each node's *bridgeness* (Nepusz et al., 2008), namely, the ability to participate across multiple communities and "bridge" them together. Bridge regions are potentially important because they have the ability to influence multiple communities.

To help characterize a node's functional role, we employed a measure of *membership diversity* capturing the extent to which a node participates in multiple communities (Najafi et al., 2016). Our reasoning was that nodes with high membership diversity may function as bridges and facilitate communication across multiple communities. Diversity was determined by computing Shannon entropy, which is used across several disciplines (e.g., biology, economics) as a measure of "diversity." Formally,

$$H_i = -\sum_{j=1}^{D} p_{i,j} \times \ln\left(p_{i,j}\right)$$

where H_i is the diversity of the i-th region, $p_{i,j}$ is the probability that the i-th region belongs to the j-th community, and D is the number of communities (ln is the natural logarithm). Thus, maximal membership diversity occurs when the region belongs equally to all communities. Because each region has membership values $\pi_{i,j}$ that take on values between 0 and 1, and membership values for each region sum to 1, we can use the $\pi_{i,j}$ as "probabilities" $p_{i,j}$ in the previous formula. To characterize node function, we also considered *degree*, which measures the extent of node connectivity. Note that *degree* and *membership diversity* capture different aspects of node function, as *degree* helps measure the extent to which a region is a hub, and *membership diversity* indicates the extent to which a region is a cross-community bridge. Combining these two measures leads to four general region classes:

- *Locally connected regions* (low degree/low diversity) are not highly connected and communicate primarily within a single community;
- *Local hubs* (high degree/low diversity) are highly connected regions that communicate primarily within a community;
- *Bottleneck bridges* (low degree/high diversity) are regions with few connections that span multiple communities;
- *Hub bridges* (high degree/high diversity) are highly connected regions with connections that span multiple communities.

Based on these considerations, we defined continuous bridgeness scores for bottlenecks and hubs. At rest, multiple *bottleneck bridges* were found in the PFC (Fig. 4), including in dorsolateral and more inferior PFC. At rest, *hub bridges* were not prominently found in the PFC (Fig. 5). Notable changes were observed during task execution. During the working memory task, several regions in the occipital cortex showed high *hub bridge* scores. Whereas the same regions were also well connected at rest (they behaved as *local hubs*, which are highly connected regions that communicate primarily within a community), and they diversified their participation across communities during the working memory task (and to some extent during the emotion task), thus increasing *bridgeness*. This is interesting in light of the fact that the task required participants to hold in mind information about multiple types of visual stimuli (places, tools, faces, and body parts), and suggests that working memory performance is characterized by the participation of visual cortex in multiple large-scale networks (see Sreenivasan et al., 2014). This is also evidenced by the stronger *hub bridges* observed in ventral temporal cortex (which were not prominent during rest).

During the emotion task, *hub bridges* in parietal cortex were stronger more inferiorly in the vicinity of the angular gyrus. Furthermore, strong *bottleneck bridges* were not prominent in dorsal prefrontal regions and were found instead more inferiorly (especially on the right hemisphere). These findings resonate with the roles attributed to the angular gyrus (Seghier, 2013) and inferior frontal regions. Interestingly, sites in the inferior frontal gyrus have been consistently reported in many emotion tasks and implicated in "emotional salience," and a recent metaanalysis has identified the

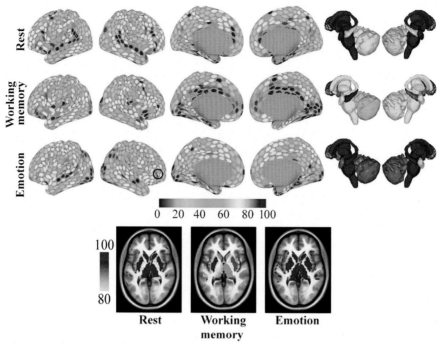

FIG. 4

Bottleneck bridges (regions with relatively few connections but that span multiple communities). Bridgeness scores for each region and condition. *Colors* indicate the percentile of the ROI's median score across 5000 bootstrapped iterations (e.g., regions colored *red* had bridgeness scores around the 90th percentile or above). *Black contours* indicate regions discussed in the text.

Reproduced with permission from Najafi, M., Mcmenamin, B.W., Pessoa, L. 2016. Overlapping communities reveal rich structure in large-scale brain networks during rest and task conditions. NeuroImage 135, 92–106.

inferior frontal gyrus (on the right hemisphere) as a hub for emotional processing (Kirby and Robinson, 2015). Our results suggest that this region may function as a *bottleneck bridge*, that is, a region that is not necessarily highly connected but one that participates across several communities. In addition, the anterior insula behaved as a strong *bottleneck bridge*, a role that was also observed at rest (but not during working memory). Finally, we note that, in our framework, nodes typically associated with the default network did not have high *bridgeness* scores. This is in contrast to reports based on *degree* that suggest that they are "globally" connected regions (e.g., Cole et al., 2010; Tomasi and Volkow, 2011). Taken together, our results demonstrate, once again, that several properties observed at rest were altered during task execution.

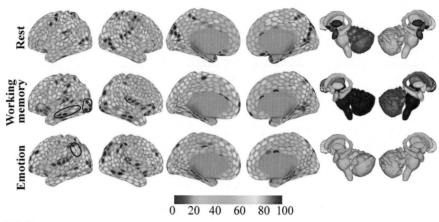

FIG. 5

Hub bridges (regions that are highly connected and have connections to multiple communities). Bridgeness scores for each region and condition. Colors indicate the percentile of the ROI's median score across 5000 bootstrapped iterations (e.g., regions colored red had bridgeness scores around the 90th percentile or above). Black contours indicate regions discussed in the text.

Reproduced with permission from Najafi, M., Mcmenamin, B.W., Pessoa, L., 2016. Overlapping communities reveal rich structure in large-scale brain networks during rest and task conditions. NeuroImage 135, 92–106.

BRAIN NETWORKS ARE DYNAMIC

The overlapping-networks framework suggests that brain regions affiliate with multiple communities. In addition, it is proposed that networks be viewed as inherently dynamic. Thus, structure-function mappings are not static but temporally varying (Pessoa, 2014). There are two important ways in which brain networks are dynamic. First, we can consider how specific networks (say, the salience network) evolve across time (Fig. 6A and B). However, this assumes that networks are stable units across time, that is, their nodes remain fixed. Second, and more generally, networks are suggested to be dynamic coalitions of brain regions that form and dissolve to meet specific computational needs. Accordingly, network descriptions need to specify how groupings of regions evolve temporally (Fig. 7). We discuss these two aspects in turn.

EVOLUTION OF NETWORK ORGANIZATION ACROSS TIME

At the spatiotemporal resolution of fMRI, we and others have started to characterize how emotion influences the temporal unfolding of large-scale network organization (Hermans et al., 2011, 2014; McMenamin et al., 2014; McMenamin and Pessoa, 2015). In one study, we focused on three networks that have been extensively studied

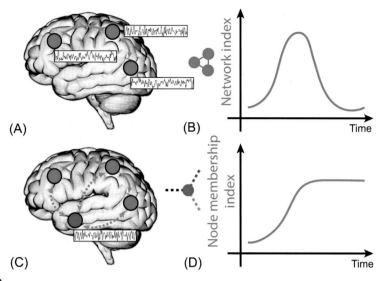

FIG. 6

Brain networks and dynamics. (A, B) Specific network properties ("network index") evolve across time. (C, D) A region's grouping with multiple networks evolves across time as indicated by the "membership index" (inset: *purple region* and its functional connections to multiple networks). The region indicated in *purple* increases its coupling with one of the networks/communities and stays coupled with it for the remainder of the time.

Reproduced with permission from Pessoa, L., 2018. Understanding emotion with brain networks. Curr. Opin.

Behav. Sci. 19, 19–25.

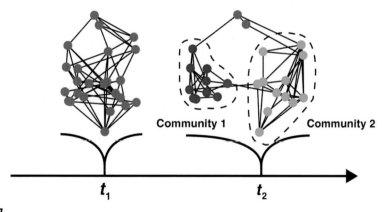

FIG. 7

Dynamic community formation. At t_1, the network is not naturally organized into clusters. At t_2, the nodes cluster into two relatively independent clusters (communities 1 and 2).

Reproduced with permission from Pessoa, L., McMenamin, B., 2016. Dynamic networks in the emotional

brain. Neuroscientist.

in the literature but whose changes during emotional manipulations is poorly understood: the salience network, the executive control network, and the task-negative network (McMenamin et al., 2014). Based on a prior study, we knew that the anticipation of mild shock altered interactions within and between networks (Kinnison et al., 2012). However, the use of short-duration trials did not allow us to characterize temporally *sustained* changes to functional connectivity. Thus, in the more recent study, the goal was to unravel changes to network organization as participants encountered a threat and continued in a "threat state" for a more prolonged duration.

Nonhuman research suggests that the amygdala plays a role in transient acute threat ("fear") responses, but the bed nucleus of the stria terminalis (BST) is implicated in more sustained potential threat ("anxious") processing (Shi and Davis, 1999; Walker et al., 2003). Correspondingly, human neuroimaging studies have found transient amygdala responses following cues of imminent threat and sustained responses during prolonged threat in locations consistent with the BST (Mobbs et al., 2010; Somerville et al., 2010; Alvarez et al., 2011). However, how the amygdala and BST interact with *other* brain networks during the processing of extended threat is poorly understood. Thus, an important goal of our study was to characterize the contributions of the amygdala and BST to network reorganization during potential threat.

In the experiment, we informed participants that a colored (say, yellow) circle on the screen indicated that they were in a threat block and mild electric shocks would be delivered randomly to their left hand, whereas a circle of another colour (say, blue) indicated that they were in a safe block and no shocks would be delivered (blocks lasted 60 s on average). Here, we discuss changes in network organization during an "early" period (transient response that peaked at 5 s after block onset) and an "intermediate" period (more sustained-response period from 7.5–17.5 s after block onset). Two network properties were investigated, *efficiency* and *betweenness*. Briefly, *efficiency* measures the strength of connections between nodes to assess the extent to which nodes can potentially influence one another; *betweenness* identifies nodes that have multiple/strong paths going through them, which may therefore be important for signal communication.

The onset of a threat period increased *efficiency* within the salience network and decreased the *betweenness* of the executive control network (Fig. 8). This suggests that threat possibly increased communication within the salience network and decreased the extent to which the executive control network facilitated communication between other networks. Put another way, signals within the salience network became more cohesive, and the executive network became more segregated from communication between other networks. Notably, although changes in network organization were detected at the onset of threat blocks, only minimal *activation* differences were observed between the threat and safe conditions. A potential reason for the lack of robust response differences is that both block onsets were motivationally salient—the onset of threat blocks because they signaled potential shock and the onset of safe blocks because they signaled safety. More generally, this example illustrates an important principle, namely, that differences in activation can be *dissociated* from differences in *co*-activation.

Effects of threat on network structure

Edge weight depicts functional connectivity between nodes
Node size depicts each subnetwork's *betweenness*

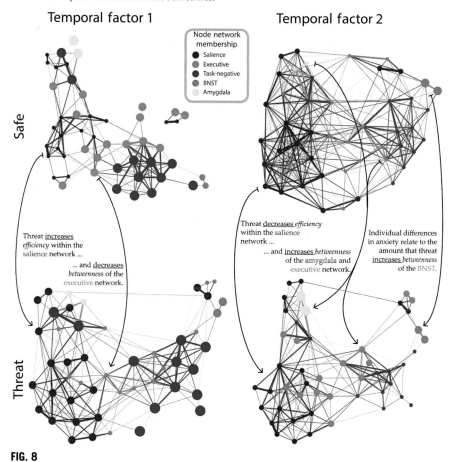

FIG. 8

Network organization as a function of threat and time. Force-layout depiction of the network organization in each condition (safe, threat) for the initial and intermediate periods. Nodes are colored according to their network affiliation, and edge width corresponds to the magnitude of the functional connectivity. Node size depicts the betweenness centrality for that network, or amygdala/BST.

Reproduced with permission from McMenamin, B.W., Langeslag, S.J., Sirbu, M., Padmala, S., Pessoa, L., 2014. Network organization unfolds over time during periods of anxious anticipation. J. Neurosci. 34, 11261–11273.

The main goal of the study was to characterize changes in network organization during sustained periods of threat monitoring ("anxious apprehension"). How were the networks organized after the initial transient processing of the threat cue? During the intermediate period, threat decreased *efficiency* in the salience network (opposite to the effect detected during the initial period), highlighting the need to characterize how network organization unfolds across time. During this period, *efficiency* also decreased in the executive and task-negative networks. In addition, threat decreased the between-network *efficiency* for all networks. In other words, during sustained threat, signals within large-scale brain networks became less coordinated, and the potential for communication when considering pairs of networks decreased.

Importantly, changes in network organization during the intermediate period involved both amygdala and BST. Threat increased amygdala *betweenness*, which is noteworthy because, in human studies, the amygdala typically does not evoke robust responses when threat is more diffuse, distal, or unpredictable. The changes in functional connectivity are consistent with the idea that, during prolonged aversive states, the amygdala influences the *communication* between brain networks more so than simply responding more strongly to threat stimuli—again highlighting the dissociation between activation and coactivation. In addition, during the same intermediate period of threat blocks, BST *betweenness* was associated with individual differences in anxiety scores, with greater increases in *betweenness* observed for participants with higher anxiety. Because the BST also exhibits responses that are more sustained during anxious states, it appears that the BST not only influences processing by generating stronger responses but becomes more central to signal communication too.

The example of changes of network organization from a transient encounter with a threat cue to a more sustained state of threat monitoring (Fig. 8) is, of course, an extremely rudimentary case of network dynamics. A more desirable description would be to track how a network property P changes with time. Thus, more generally, we would like to estimate $P(N,t)$, that is, the property P at time t for network N (if the property in question is at the network level) or node N (if the property in question is at the node level). Indeed, the past few years have witnessed the fast growth of techniques to study the dynamics of networks in general (e.g., Kolar et al., 2010; Boccaletti et al., 2014) and brain networks in particular (e.g., Brauna et al., 2015).

FLUID NETWORK IDENTITY ACROSS TIME

The discussion in the previous section assumed that networks are stable units across time, that is, their nodes remain fixed. But, what if across time some nodes get added or deleted from a community? In this scenario, network descriptions need to specify how groupings of regions evolve temporally (Fig. 7). More generally, networks are suggested to be dynamic coalitions of brain regions that form and dissolve to meet specific computational needs. To be true, this type of approach poses several challenges as the notion of a network as a coherent unit is undermined. For instance, at what point does a coalition of regions become something *other* than, say, the salience network?

An illustration of the previous posed questions is provided by a study by Hearne et al. (2017). During the experimental session, participants performed a series of cognitive tasks of varying difficulty levels (each of which was blocked). The four cognitive conditions were preceded and followed by resting periods, totaling six conditions/periods. Standard (disjoint) community detection was performed on the first resting-state period, during which four communities were detected (visual, sensory, frontoparietal, default). In other words, a subset of the brains' Region of Interests (ROIs) were assigned to the visual community, another subset assigned to the sensory community, and so on. Community detection was also performed for all remaining five conditions/periods. If we consider a specific ROI, we can trace the communities it belongs to by drawing a line that links the ROI across each of the six conditions/periods, and when this is done for all ROIs, we have a so-called alluvial plot (Fig. 9).

Let's consider the first two periods, resting state and "null" (during which participants passively looked at the cognitive problems). Community detection during "null" also revealed four communities. However, some of the specific node assignments changed. For example, some nodes in the frontoparietal community at rest were assigned to the sensory community during "null" (see the green lines from the frontoparietal community at rest to the sensory community during "null"). But,

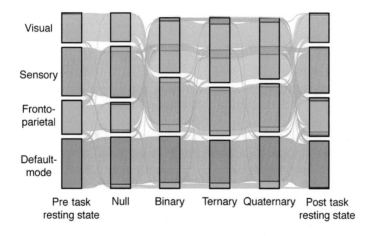

FIG. 9

Network organization and task condition. Community structure for each of six experimental tasks/conditions. ROIs are arranged vertically and assigned to three or four communities. Each line represents an ROI and the trace across conditions indicates the community assignment of the ROI. For example, for the pretask resting state, ROIs belonging to the visual network are colored in blue. The *color* of each line identifies the original community assignment during the pretask resting state. For further information on alluvial plots, see https://en.wikipedia.org/wiki/Alluvial_diagram.

Reproduced with permission from Hearne, L.J., Cocchi, L., Zalesky, A., Mattingley, J.B., 2017. Reconfiguration of brain network architectures between resting-state and complexity-dependent cognitive reasoning. J. Neurosci. 37, 8399–8411.

although some nodes switched affiliation, by and large, the communities were composed of the same nodes. A rather different picture emerged during the "binary" task, during which only three communities were detected. In particular, the first community (top) was comprised of nodes that originally were assigned to the visual (blue), frontoparietal (green), and sensory (brown) communities. So we can see that, during the "binary" task, the nodes rearranged themselves into different groupings (i.e., communities). A few changes were observed during the "ternary" and "quaternary" conditions, but the overall organization across the three communities was relatively stable. The final transition between the "quaternary" condition and the final resting state shows how the four typical communities emerged again (although minor differences can be identified relative to the partition during the first rest period).

The example illustrates how we can think of networks assembling and dissolving (Fig. 7). At a given time, the organization of the system may be well described by four communities but later better described in terms of three communities (to stay close to the example discussed here). From the perspective of a node, it will cohere with a given network at one time, then flexibly affiliate with another community. In many instances, subcollections of nodes will form natural clusters that jointly affiliate with larger communities, such as the nodes of the frontoparietal network in Fig. 9 (one can mentally trace the "green stream" along the experimental conditions).

In the ongoing discussion, because disjoint networks were considered, nodes were assigned to a single community. The mixed-membership framework discussed in the first part of the chapter offers a powerful alternative, where membership is not binary and instead varies continuously between 0 and 1. To incorporate dynamics, each node can be a member of multiple networks with a set of probability-like membership values, and these can fluctuate across time (Fig. 6C and D). The advantage of having continuous-valued membership values is that a node is allowed to participate in more than one community, which can possibly provide a richer characterization of the overall network organization. For example, in the context of the study by Hearne et al. (2017), some of the nodes in the sensory community at rest that are part of the first (top) community during the "binary" task may also participate in the second community (middle) again during the "binary" task, which is mostly comprised of nodes originally in the sensory community. Naturally, if the data are most consistent with a fairly disjoint network structure, membership values will tend be close to 1 for only one community. In effect, the mixed-membership is a more general framework that subsumes the disjoint approach typically utilized.

CONCLUSIONS

Where are functions instantiated in the brain? This is the question that was posed at the beginning of the chapter. Increasingly, the understanding of structure-function mappings in the brain is conceptualized in terms of distributed circuits (e.g., Pessoa, 2017). The development of network science in the past 20 years (Newman, 2010) has provided a renewed impetus to this approach. Particularly interesting is the fact that human neuroimaging, which for a decade and a half was often (at least informally) described as neophrenology, has strongly embraced a more distributed framework. This is an illuminating example of how the availability of tools can have such a drastic effect on conceptual frameworks applied to data.

The present chapter discussed two conceptual aspects to understanding brain networks, namely, overlap and dynamics. Currently, a broad range of network techniques exist that allow investigators to address these themes head on, and this will contribute to refining and extending the understanding of how complex distributed computations bring about complex mental function.

REFERENCES

Airoldi, E.M., Blei, D.M., Fienberg, S.E., Xing, E.P., 2008. Mixed membership stochastic blockmodels. J. Mach. Learn. Res. 9, 1981–2014.

Alvarez, R.P., Chen, G., Bodurka, J., Kaplan, R., Grillon, C., 2011. Phasic and sustained fear in humans elicits distinct patterns of brain activity. Neuroimage 55, 389–400.

Boccaletti, S., Bianconi, G., Criado, R., Del Genio, C.I., Gómez-Gardenes, J., Romance, M., Sendina-Nadal, I., Wang, Z., Zanin, M., 2014. The structure and dynamics of multilayer networks. Phys. Rep. 544, 1–122.

Brauna, U., Schäfera, A., Walterb, H., Erkb, S., Romanczuk-Seiferthb, N., Haddada, L., Schweigera, J.I., Grimma, O., Heinzb, A., Tosta, H., Meyer-Lindenberga, A., Bassettc, D.S., 2015. Dynamic reconfiguration of frontal brain networks during executive cognition in humans. Proc. Natl. Acad. Sci. 112, 11678–11683.

Bressler, S.L., Menon, V., 2010. Large-scale brain networks in cognition: emerging methods and principles. Trends Cogn. Sci. 14, 277–290.

Cocchi, L., Zalesky, A., Fornito, A., Mattingley, J.B., 2013. Dynamic cooperation and competition between brain systems during cognitive control. Trends Cogn. Sci. 17, 493–501.

Cole, M.W., Pathak, S., Schneider, W., 2010. Identifying the brain's most globally connected regions. Neuroimage 49, 3132–3148.

Cole, M.W., Reynolds, J.R., Power, J.D., Repovs, G., Anticevic, A., Braver, T.S., 2013. Multi-task connectivity reveals flexible hubs for adaptive task control. Nat. Neurosci. 16, 1348–1355.

Evans, T.S., Lambiotte, R., 2009. Line graphs, link partitions, and overlapping communities. Phys. Rev. E Stat. Nonlinear Soft Matter Phys. 80, 016105.

Finger, S., 1994. Origins of Neuroscience: A History of Explorations into Brain Function. Oxford University Press, New York.

Fornito, A., Zalesky, A., Bullmore, E., 2016. Fundamentals of Brain Network Analysis. Academic Press, San Diego.

Gavin, A.-C., Bösche, M., Krause, R., Grandi, P., Marzioch, M., Bauer, A., Schultz, J., Rick, J.M., Michon, A.-M., Cruciat, C.-M., 2002. Functional organization of the yeast proteome by systematic analysis of protein complexes. Nature 415, 141–147.

Gopalan, P., Blei, D., 2013. Efficient discovery of overlapping communities in massive networks. Proc. Natl. Acad. Sci. 110, 14534–14539.

Guimera, R., Nunes Amaral, L.A., 2005. Functional cartography of complex metabolic networks. Nature 433, 895–900.

Hearne, L.J., Cocchi, L., Zalesky, A., Mattingley, J.B., 2017. Reconfiguration of brain network architectures between resting-state and complexity-dependent cognitive reasoning. J. Neurosci. 37, 8399–8411.

Hermans, E.J., van Marle, H.J., Ossewaarde, L., Henckens, M.J., Qin, S., van Kesteren, M.T., Schoots, V.C., Cousijn, H., Rijpkema, M., Oostenveld, R., Fernandez, G., 2011. Stress-related noradrenergic activity prompts large-scale neural network reconfiguration. Science 334, 1151–1153.

Hermans, E.J., Henckens, M.J., Joëls, M., Fernández, G., 2014. Dynamic adaptation of large-scale brain networks in response to acute stressors. Trends Neurosci. 37, 304–314.

Hoffman, M.D., Blei, D.M., Wang, C., Paisley, J., 2013. Stochastic variational inference. J. Mach. Learn. Res. 14, 1303–1347.

Kinnison, J., Padmala, S., Choi, J.M., Pessoa, L., 2012. Network analysis reveals increased integration during emotional and motivational processing. J. Neurosci. 32, 8361–8372.

Kirby, L.A., Robinson, J.L., 2015. Affective mapping: an activation likelihood estimation (ALE) meta-analysis. Brain Cogn. 118, 137–148.

Kolar, M., Song, L., Ahmed, A., Xing, E.P., 2010. Estimating time-varying networks. Annal. Appl. Stat. 4, 94–123.

McIntosh, A.R., 2000. Towards a network theory of cognition. Neural Netw. 13, 861–870.

McMenamin, B.W., Pessoa, L., 2015. Discovering networks altered by potential threat ("anxiety") using quadratic discriminant analysis. Neuroimage 116, 1–9.

McMenamin, B.W., Langeslag, S.J., Sirbu, M., Padmala, S., Pessoa, L., 2014. Network organization unfolds over time during periods of anxious anticipation. J. Neurosci. 34, 11261–11273.

Menon, V., Uddin, L.Q., 2010. Saliency, switching, attention and control: a network model of insula function. Brain Struct. Funct. 214, 655–667.

Mesulam, M.M., 1990. Large-scale neurocognitive networks and distributed processing for attention, language, and memory. Ann. Neurol. 28, 597–613.

Mobbs, D., Yu, R., Rowe, J.B., Eich, H., FeldmanHall, O., Dalgleish, T., 2010. Neural activity associated with monitoring the oscillating threat value of a tarantula. Proc. Natl. Acad. Sci. U. S. A. 107, 20582–20586.

Najafi, M., Mcmenamin, B.W., Pessoa, L., 2016. Overlapping communities reveal rich structure in large-scale brain networks during rest and task conditions. Neuroimage 135, 92–106.

Nepusz, T., Petróczi, A., Négyessy, L., Bazsó, F., 2008. Fuzzy communities and the concept of bridgeness in complex networks. Phys. Rev. E 77, 016107.

Newman, M., 2010. Networks: An Introduction. Oxford University Press, New York City.

Palla, G., Derenyi, I., Farkas, I., Vicsek, T., 2005. Uncovering the overlapping community structure of complex networks in nature and society. Nature 435, 814–818.

Pessoa, L., 2013. The Cognitive-Emotional Brain: From Interactions to Integration. MIT Press, Cambridge.

Pessoa, L., 2014. Understanding brain networks and brain organization. Phys. Life Rev. 11, 400–435.

Pessoa, L., 2017. A network model of the emotional brain. Trends Cogn. Sci. 21, 357–371.

Seghier, M.L., 2013. The angular gyrus: multiple functions and multiple subdivisions. Neuroscientist 19, 43–61.

Shi, C., Davis, M., 1999. Pain pathways involved in fear conditioning measured with fear-potentiated startle: lesion studies. J. Neurosci. 19, 420–430.

Somerville, L.H., Whalen, P.J., Kelley, W.M., 2010. Human bed nucleus of the stria terminalis indexes hypervigilant threat monitoring. Biol. Psychiatry 68, 416–424.

Sporns, O., 2010. Networks of the Brain. MIT Press, Cambridge, MA.

Sreenivasan, K.K., Curtis, C.E., D'Esposito, M., 2014. Revisiting the role of persistent neural activity during working memory. Trends Cogn. Sci. 18, 82–89.

Tomasi, D., Volkow, N.D., 2011. Functional connectivity hubs in the human brain. Neuroimage 57, 908–917.

Walker, D.L., Toufexis, D.J., Davis, M., 2003. Role of the bed nucleus of the stria terminalis versus the amygdala in fear, stress, and anxiety. Eur. J. Pharmacol. 463, 199–216.

Wasserman, S., Faust, K., 1994. Social Network Analysis: Methods and Applications. Cambridge University Press, New York.

Xie, J., Kelley, S., Szymanski, B.K., 2013. Overlapping community detection in networks: the state-of-the-art and comparative study. ACM Comput. Surv. 45, 43.

Yeo, B.T., Krienen, F.M., Sepulcre, J., Sabuncu, M.R., Lashkari, D., Hollinshead, M., Roffman, J.L., Smoller, J.W., Zollei, L., Polimeni, J.R., Fischl, B., Liu, H., Buckner, R.L., 2011. The organization of the human cerebral cortex estimated by intrinsic functional connectivity. J. Neurophysiol. 106, 1125–1165.

FURTHER READING

Pessoa, L., 2018. Understanding emotion with brain networks. Curr. Opin. Behav. Sci. 19, 19–25.

Pessoa, L., McMenamin, B., 2017. Dynamic networks in the emotional brain. Neuroscientist 23, 383–396.

The uniqueness of the individual functional connectome

4

Corey Horien*, Dustin Scheinost[†,‡], R. Todd Constable*[,†,§]

Interdepartmental Neuroscience Program, Yale University School of Medicine, New Haven, CT, United States Department of Radiology and Biomedical Imaging, Yale University School of Medicine, New Haven, CT, United States[†] The Child Study Center, Yale University School of Medicine, New Haven, CT, United States[‡] Department of Neurosurgery, Yale University School of Medicine, New Haven, CT, United States[§]*

CHAPTER OUTLINE

INTRODUCTION

Humans exhibit significant variation in genotypic, phenotypic, and behavioral characteristics (Collins et al., 1998; Finn et al., 2015; Frazer et al., 2009; Hariri et al., 2002; Heils et al., 1996; Hines, 2010; Iafrate et al., 2004; Mueller et al., 2013; Nettle, 2006; Rahim et al., 2008; Redon et al., 2006). A goal of modern neuroscience is to better understand underlying individual differences at multiple spatial and temporal scales using a variety of methodological approaches, ranging from molecular tools to study (epi)genetic differences (Hannon et al., 2015; Houston et al., 2013; Illingworth et al., 2015; Sweatt and Tamminga, 2016) to exploring how brain patterns collected in neuroimaging studies relate to differences in cognitive and clinical data (Finn et al., 2015; Hearne et al., 2016; Kaufmann et al., 2017; Poole et al., 2016; Rosenberg et al., 2016a, b, 2018; Tavor et al., 2016). Despite the widespread acknowledgment of interindividual heterogeneity, studies in human neuroimaging,

Connectomics. https://doi.org/10.1016/B978-0-12-813838-0.00004-2

including functional MRI (fMRI), typically involve separating subjects into groups based on clinical or behavioral measures and investigating group-level differences in brain function (Blumberg et al., 2003; Donegan et al., 2003; Durston et al., 2003; Finn and Constable, 2016; Gee et al., 2013, 2014; Hare et al., 2008; Li et al., 2005; Potenza et al., 2003; Shaywitz et al., 2002). Such studies have significantly advanced understanding of brain structure and function in both health and disease.

Nevertheless, there is increasing interest in studying individual subjects using fMRI. By utilizing an individualized approach to data analysis, investigators are increasingly able to better characterize unique patterns of brain connectivity and generate meaningful models about cognitive, behavioral, or clinical features. Importantly, characterizing individual differences has been noted to afford researchers the opportunity to predict relevant behavioral and health-related outcomes (Gabrieli et al., 2015). We recently demonstrated that individual functional connectomes are unique, and that differences across individuals are meaningful and can be related to behavior (Finn et al., 2015). The uniqueness of individual connectome data was revealed by our ability to identify (ID) individual subjects from a collection of such data from many subjects. Although there are many ways to ID individuals and the goal of fMRI is not to provide yet another means of doing so, measuring connectome-based identification rates is a useful approach for highlighting individual differences in the functional organization of the brain. Consideration of identification rates can help to reveal the features that drive individual differences or the factors that enhance such differences. Understanding these factors can lead to a better prediction framework for building models relating brain organization to behavior or clinical symptoms (Finn et al., 2015). Functional connectome analysis allows investigators to study whole-brain patterns of organization as well as network- and node-level properties in individual subjects. In this chapter, we describe the methodology for assessing identification using fMRI functional connectivity (FC) data and discuss results to date illustrating connectome stability in different populations, the brain regions that make individuals unique, and factors influencing individual differences, including brain state manipulations through task-based scans. These discussions are presented as a means for informing studies related to predicting brain-behavior relationships through connectome-based predictive modeling methods (Finn et al., 2015, Shen et al., 2017).

IDENTIFICATION PROCEDURE

Connectome-based identification uses FC data from fMRI studies to generate predictions about a subject's identity from a group (Finn et al., 2015; see reference for a detailed description of the method). For example, consider a dataset consisting of n subjects with two resting-state scans acquired on separate days. A database is first created consisting of all the subjects' connectivity matrices from the second session (Fig. 1). A connectivity matrix from a particular subject is then selected from the first session and denoted as the target. Pearson correlation coefficients are then computed between the target connectivity matrix and all the matrices in the database

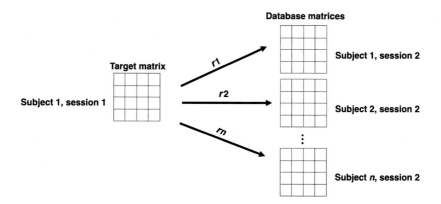

Predicted identity = argmax(r1, r2.....rn)

FIG. 1

The identification procedure. Pearson correlation coefficients are calculated between the target matrix and all database matrices. The predicted identity is the target-database pair resulting in the highest correlation coefficient.

from session 2. If the highest Pearson correlation coefficient was between subject 1, session 1 and subject 1, session 2, this would be recorded as a correct identification. In essence, the method is based on intra- and intersubject correlations: if a participant's intrasubject correlation is greater than the same participant's intersubject correlation with every other participant (Finn et al., 2015), a successful identification is recorded. Subject 2 from session 1 would now be denoted as the target, and the algorithm would continue with the same database of subjects. The process is repeated until identifications have been performed for all session and database-target combinations. Note that the method is flexible such that it allows the use of whole-brain connectivity data or connectivity data from specific networks to be utilized. Statistical significance can be determined using permutation testing, in which a null distribution is created by randomly shuffling subject identities and performing the identification algorithm with the incorrect subject labels; identification rates obtained from the original sample are then compared with the null distribution to determine if identification occurred at above chance levels. Bootstrap methods can also be used to determine confidence intervals for a given identification.

FC IDENTIFICATION: EARLY WORK AND RESULTS IN ADULT SUBJECTS

The first study describing connectome-based identification and demonstrating its efficacy (Finn et al., 2015) in a large group of subjects utilized a dataset from the Human Connectome Project (HCP; Van Essen et al., 2013). The HCP data used in the original study consisted of 126 adult subjects (ages 22–35, 40 males) scanned on two separate days; subjects completed a rest scan on each day and also engaged in various task-based scans. Identification rates ranged from 54% to 94.4% using whole brain data, which was highly above chance ($P < .0001$). Using a combination of the best performing networks, rates as high as 99% were achieved, again highly above chance ($P < .0001$). Numerous control analyses were conducted to ensure that high-within subject correlations were being driven by FC as opposed to subject-specific idiosyncrasies in head motion and anatomy, and these results confirmed that the method is capturing meaningful information. Also of note, the subjects investigated in the original study contained numerous blood relatives and sets of twins, potentially making the identification process more difficult. Nevertheless, the high ID rates achieved highlight the efficacy of the method and the extent to which it captures individual differences in FC. Thus, this work established that unique identifiable features were contained in the FC matrix to the extent that it was possible to use this information to ID adult subjects across multiple days from a pool of subjects. Since then, numerous studies have been published replicating high identification rates and the ability to reveal individual differences in adult subjects in independent datasets and in larger samples of the HCP (Finn et al., 2017; Horien et al., 2017a; Noble et al., 2017; Vanderwal et al., 2017; Waller et al., 2017). Although multiple factors (which we consider later in the section titled "Factors Affecting the Detection of Individual Differences") may influence the extent to which functional connectomes are unique and identifiable, it is of note that all studies published to date have reported accuracies highly above chance. Thus, adult FC patterns across the entire brain appear to be unique and stable on a day-to-day basis, and the identification method is an effective means to assess differences in FC data on the single-subject level.

RESULTS IN ADOLESCENTS AND APPLICATIONS TO DISEASE

Connectome-based identification has also been demonstrated in children and adolescent subjects, highlighting the fact that individual differences in FC are present in younger participants. Kaufmann et al. (2017) tested the procedure in 797 subjects ages 8–22 from the Philadelphia Neurodevelopmental Cohort (PNC; Satterthwaite et al., 2016) who completed a working-memory task, an emotion recognition task, and a resting-state scan during the same day. Using whole-brain data, the authors achieved identification rates ranging from 37%–57%, with all results highly above chance ($P < .0001$), illustrating that FC differences seem to appear in subjects as young as 8 years old. Follow-up studies in three independent datasets (Fig. 2; Horien et al., 2017b) have also demonstrated individual differences in FC patterns are detectable at rates well above chance ($P < .0001$). High rates of identification were possible despite the fact that subjects in all three samples had been scanned years apart, confirming for the first time the individual functional connectome appears to be stable over longer time frames. Further, subject discriminability did not appear to be driven by subject-specific head movements (a concern we detail in "Factors Affecting the Detection of Individual Differences"; Fig. 2). The high rates of identification are intriguing when one considers the developmental changes in brain structure and function that adolescents and young adults experience (Brenhouse and Andersen, 2011;

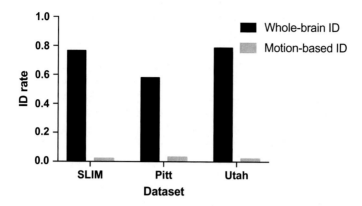

FIG. 2

Identification is possible in adolescents and young adults scanned 1–3 years apart, and results are not driven by head motion. Identification rate is indicated on the y-axis. Three independent datasets are shown under the x-axis. SLIM refers to the Southwest University Longitudinal Imaging Multimodal dataset (Liu et al., 2017); $n = 105$; average subject age = 19.7 +/−0.9 years; average time between scans = 1.58 +/−0.25 years. Pitt refers to data collected at the University of Pittsburgh; $n = 93$; average subject age = 15.2 +/−2.8 years; average time between scans = 2.15 +/−0.33 years. Utah refers to data collected at the University of Utah; $n = 26$; average subject age = 20.2 +/−8.3 years; average time between scans = 2.5 +/−0.3 years. Shown in *black* are identification rates using whole-brain data. Shown in *gray* are results of control analyses conducted using a characteristic motion vector capturing subject-specific head movements in the scanner. Identification rates based on the motion vector are much lower than identification rates achieved using whole-brain FC data, indicating that detection of individual differences is possible in developing subjects, and confounders related to motion are unlikely to drive the effect. Note that Pitt and Utah data were downloaded from the Consortium for Reliability and Reproducibility (CoRR; Zuo et al., 2014) website (http://fcon_1000.projects.nitrc.org/indi/CoRR/html/).

Burnett et al., 2011; Casey et al., 2000, 2008, 2010; Casey and Jones, 2010; Giedd et al., 2009; Giedd and Rapoport, 2010; Gogtay et al., 2004; Johnson et al., 2009; Tamnes et al., 2013; Tau and Peterson, 2010), which should make identification harder. That identification rates are still highly significant even in adolescents suggests that there is an intrinsic stability in FC patterns. Together with the data of Kaufmann et al. (2017), these results suggest children and adolescents have a unique FC profile as indicated by the high identification rates obtained with data spanning up to 3 years.

Although most studies to date have been conducted in healthy participants, clues are beginning to emerge about the relevance of these unique FC patterns to different disease states. Kaufmann et al. (2017) used the connectome-based identification method in the PNC dataset to derive a measure they termed "connectome distinctiveness", which they computed as the average identification accuracy across all identification pairs. Specifically, the authors first established that connectome distinctiveness increases with age in children and adolescents—the older the subjects were in their study, the more often they tended to be successfully identified. The authors went on to show that participants with increased preclinical symptom scores tended to exhibit a pronounced delay in connectome distinctiveness compared with healthy subjects; the functional connectomes of these individuals tended to vary more from scan-to-scan. In particular, lower connectome distinctiveness was significantly correlated with higher scores on measures of inattention, prodromal symptoms of schizophrenia, and depression. Though this study was cross-sectional in nature, the results indicate that the method can be applied to draw inferences about symptom scores and disease-specific features.

COMMON ANATOMICAL THEMES

In all of the studies performed to date, there are clear differences in terms of which brain regions and networks are helpful in identification and those that are not. Networks comprised of regions of interest (ROIs; "nodes" in graph theory) in the medial frontal, frontoparietal, and to a lesser extent the default mode network tend to contribute substantially to a successful identification and hence underlie much of the individual variability in FC data. For example, if an identification is restricted to using only nodes within the medial frontal or frontoparietal regions (or a combination of these two networks), identification rates approach or even surpass rates obtained using whole-brain data (Finn et al., 2015, 2017; Horien et al., 2017b; Vanderwal et al., 2017; Waller et al., 2017). Between-network connections involving these highly distinctive regions also tend to lead to high identification rates (Horien et al., 2017b). Conversely, within- and between-network connections in the subcortical-cerebellum, sensorimotor, and visual areas tend to be less distinctive, though identification rates are still usually well above chance. Additional analyses calculating the contribution of specific connections ("edges" in graph theory) to a successful identification reinforce that edges in the frontal, parietal, and temporal association cortices tend to contribute to a successful identification (Finn et al., 2015; Horien et al., 2017b; Vanderwal et al., 2017). These data indicate that FC patterns in higher-order association cortices tend to be distinctive within an individual, whereas primary sensory and motor regions tend to exhibit similar patterns between individuals (Finn et al., 2015). The fact that networks in the frontoparietal regions are most evolutionarily recent (Zilles et al., 1988) and hence underlie many unique aspects of cognition has been posited as one of the reasons for the high interindividual variance underlying detection of individual differences (Finn et al., 2015). Similarly, the high interindividual variance in higher-order association cortices might underlie the low pattern of group-level reproducibility observed in these regions in other work, whereas less distinctive areas like the sensorimotor strip tend to show higher group-level reproducibility (Shen et al., 2013). In addition, it is interesting to note that Kaufmann et al. (2017) found that connectome-distinctiveness in the highly unique regions—specifically the medial frontal, frontoparietal, and default mode networks—tended to drop off in individuals with higher psychiatric symptom scores. These results underscore the importance of these networks in normal health, while highlighting the extent to which they might be vulnerable to, and ultimately cause, certain psychiatric disorders.

FACTORS AFFECTING THE DETECTION OF INDIVIDUAL DIFFERENCES

In this section, we discuss numerous factors affecting the extent to which individual differences in FC data can be detected through identification, including subject motion, amount of scan time, acquisition sequence, parcellation scheme, and participant gender. Due to the myriad ways in which preprocessing strategies can differ, a full consideration is outside the scope of this chapter. However, we note that the studies published to date have relied on preprocessing pipelines differing in nontrivial ways, and all have reported essentially consistent results—identification at levels well above chance—using the method.

Subject motion in the scanner has been shown to be a confounder in estimates of FC (Power et al., 2015; Satterthwaite et al., 2012). It is not surprising, therefore, that including high-motion subjects in identification analyses is a concern (Finn et al., 2017) and has been observed to decrease accuracy rates (Horien et al., 2017a) even after performing steps to account for motion in preprocessing. Including strict motion thresholds can help in this regard. For instance, imposing a 0.1-mm mean frame-frame displacement (FFD) threshold (i.e., only including subjects in the analysis if they had a mean FFD < 0.1 mm) resulted in increased identification rates relative to a group consisting of all subjects (Horien et al., 2017b). The results do not seem to be driven merely by a decreased sample size, as one of the studies (Horien et al., 2017a) documented higher identification rates for a low-motion group consisting of approximately 600 subjects compared with a high-motion group of approximately 200 subjects.

Given the effect of motion on FC, it is reasonable to question if subject-specific motion in the scanner is driving the detection of individual differences. To address this question, Finn et al. (2015) developed a metric computing a motion distribution vector that captures potential characteristic movement patterns for each subject in each scan, and vectors were then submitted to the identification pipeline similar to how matrices are used for identification. Identification rates were found to be approximately 2.5%, well below the rates obtained using FC data. Follow-up studies using the same metric have also demonstrated that motion-based identification rates are much lower than the rates achieved using FC data in both adults (Vanderwal et al., 2017) and adolescents (Horien et al., 2017b; Fig. 2). Further, the fact that one study (Horien et al., 2017b) documented lower identification rates among children relative to adolescents and young-adults also makes it unlikely that subject-specific movement patterns are playing a major role in successful identification, as motion is notoriously a concern in FC studies involving younger participants. Hence, it is not likely that high within-subject correlations and identification rates are due to characteristic head movements by a subject but rather to actual differences in FC.

Scan time is another variable affecting the identifiability of individual subject FC data. It has been shown numerous times that increasing the amount of total scan time can significantly affect identification rates when using whole-brain or network-specific data (Finn et al., 2015, 2017; Horien et al., 2017a; Noble et al., 2017). Specifically, there is a positive correlation between the amount of data collected per subject and the detection of individual differences, though a plateau has been noted in a few studies (Finn et al., 2015; Horien et al., 2017a; Noble et al., 2017). In a study examining the effect of time on the identifiability of HCP subjects (via truncating time courses of individual nodes), identification rates increased for every extra minute of data up until approximately 9 of the full 14 minutes of rest data had been used (Horien et al., 2017a). Notably, the leveling off of identification rates likely relates to the algorithm used in the identification process, which necessitates that a within-subject correlation for a subject needs to be greater than the maximum of the between-subject correlations with every other participant for a successful prediction to be recorded (Noble et al., 2017). This implies that a high identification rate can be achieved without an increase in reliability of the data. Thus, individual edges might change from scan-to-scan, which can affect univariate estimates of reliability (Noble et al., 2017; Pannunzi et al., 2017), but the multivariate nature of the identification algorithm captures the overall subject-specific pattern and thus allows detection of individual differences (Noble et al., 2017).

The importance of time in identification studies is consistent with other work showing that longer acquisition times, and therefore more data per subject, are associated with increases in the reliability of FC measures (Birn et al., 2013; Laumann et al., 2015; Mueller et al., 2015; Noble et al., 2017; Shah et al., 2016) in addition to estimates of individual differentiation (Airan et al., 2016). Given that the signal of interest in FC studies is in the low frequency range (0.01–0.1 Hz), longer scan durations allow more cycles of interest to be detected, hence allowing discrimination of unique, subject-specific features. If time is not taken into account when interpreting identification rates, incorrect conclusions might be drawn regarding the identifiability of specific populations, different age groups, etc. (Horien et al., 2017a). Thus total amount of data per subject is a crucial factor that can alter subject identifiability, though this does not necessarily imply high test-retest reliability.

Although the amount of data per subject has been demonstrated to affect the degree to which individual differences are apparent, the particular acquisition sequence utilized does not appear to matter to the same extent (Horien et al., 2017a). Because the original identification study was performed in the HCP dataset that utilized a multiband acquisition sequence affording high spatial and temporal resolution, there have been concerns that historical neuroimaging acquisitions (i.e., nonmultiband), with lower spatial and temporal resolution, might not allow individual differences in the connectome to be detected (Waller et al., 2017). However, work directly comparing data collected via multiband and nonmultiband sequences observed no differences in identification rates (Horien et al., 2017a). Individual differences in FC appear to be so robust, in fact, that subsampling data to achieve an effective repetition time (TR) of 7.2 seconds still provided identification accuracies of approximately 85% in one study (Horien et al., 2017a). The consistent results at different spatial and temporal resolutions are largely in line with the work of Airan et al. (2016), who found that there was no clear difference in estimates of individual differentiation in multiband versus nonmultiband protocols. Therefore, individual patterns in the connectome are sufficiently robust that differences can be observed even in the absence of high temporal and spatial resolution.

To calculate a connectivity matrix, nodes must first be defined, and there is no consensus currently on how to define nodes or which atlas to use. A number of functional parcellation algorithms have been proposed (Craddock et al., 2012; Glasser et al., 2016; Gordon et al., 2016; Schaefer et al., 2017; Shen et al., 2013; Yeo et al., 2011) and although the effects of the particular parcellation method have not been exhaustively studied in the context of identifiability, the first work describing FC-based identification (Finn et al., 2015) compared a 268 node functional atlas and network definitions generated by the authors (Finn et al., 2015) with the FreeSurfer node atlas (68 nodes; Fischl et al., 2004) and Yeo network definitions (Buckner et al., 2011). The authors found that the 268-node atlas provided higher identification rates than the FreeSurfer atlas, suggesting that higher resolution parcellations allow better detection of individual features. Vanderwal et al. (2017) also observed that higher parcellation resolution increased identification accuracy. Nevertheless, a note of caution is warranted, as it has been suggested that increasing parcellation resolution might have unintended consequences, specifically augmenting uninteresting individual differences due to registration errors, anatomical misalignment, or motion-related errors (Finn et al., 2017). Much remains to be learned about how particular parcellations affect identification results, though it is encouraging that individual FC differences can still be detected using a variety of approaches.

It is currently unclear if there are differences with respect to gender in the detection of individual differences in FC. Briefly, one study has documented no effect of gender (Horien et al., 2017b), whereas Kaufmann et al. (2017) showed a clear sexual dimorphism in identification rates and connectome distinctiveness using whole-brain data. Additionally, Finn et al. (2017) have shown gender affects between-subject correlations and effect sizes vary by scan condition. Given that there are numerous studies showing gender differences with respect to resting-state FC (Biswal et al., 2010; Kilpatrick et al., 2006; Satterthwaite et al., 2015; Scheinost et al., 2015), more work is needed to better understand how participant gender affects identifiability of the functional connectome.

IDENTIFICATION IN REST, TASK, AND NATURALISTIC SCANS

As discussed in detail in Finn et al. (2017), resting-state scans have formed the backbone of FC research due to various reasons, including convenience for the subject and ease of standardization across institutions, and it is free of many of the confounders found in task-based fMRI (Greicius, 2008; Smith et al., 2013). Although rest has been the default condition for FC research, there has been increased interest in investigating the use of task paradigms, particularly continuous performance tasks, or naturalistic conditions in which participants view complex scenes, such as from a movie (Finn et al., 2017; Vanderwal et al., 2017). Although initial work in the HCP dataset seemed to suggest that rest was the optimal condition to increase intrasubject correlations and hence identifiability from scan-to-scan (Finn et al., 2015), these results were shown to be due to the amount of data—more data per subject in the rest scans compared with task scans afforded higher identifications. Since then, it has been demonstrated that identification rates tend to be highest when subjects are engaged in tasks or in viewing naturalistic scenes in both the target and database scans (Finn et al., 2017; Vanderwal et al., 2017). (Interestingly, an analysis conducted by Finn et al. (2015) in the original identification work suggested that task-based scans might indeed allow higher identifications compared with rest. When expanding the database of matrices to include both a rest and a task scan and using a task target scan, higher identification rates were obtained compared with using only an all-rest or all-task database.) Two hypotheses have been proposed regarding individual differences in connectivity matrices acquired during task-based scans. First, tasks have been suggested to increase subject-specific signal-to-noise (Finn et al., 2017; Vanderwal et al., 2017), essentially maximizing both intrasubject correlations and making interindividual variability more easily detectable. Although this might seem counterintuitive given that all subjects in a dataset will have completed the same task (in principle activating the same brain networks across individuals and potentially making them look more similar), the within-subject correlation increases observed in tasks seem to be enough to override the resultant between-subject correlation increases (Finn et al., 2017). Most likely driving the task-induced within-subject correlation increases are individual-specific changes in the frontoparietal network (Finn et al., 2017; Vanderwal et al., 2017), a highly distinctive network at baseline (Finn et al., 2015). In other words, tasks appear to enhance the connectivity features that make each subject unique. In contrast, connectivity patterns measured during rest are relatively noisy. As detailed in Finn et al. (2017), FC patterns have been found to change as a function of arousal (Chang et al., 2016; Tagliazucchi and Laufs, 2014), mood (Harrison et al., 2008), what participants are thinking about in the scanner (Christoff et al., 2009; Gorgolewski et al., 2014; Shirer et al., 2012), or whether the participant's eyes were open or closed during scanning (Patriat et al., 2013), in addition to other influences (Barnes et al., 2009; Duncan and Northoff, 2013; Tung et al., 2013). Hence, the wider range of brain states during rest, and the effect of brain state on the FC matrix, make identification of individuals via their functional connectome more difficult with resting-state scans (Finn et al., 2017).

The second reason for the improvement in identifiability of subjects under task conditions is likely due to pragmatic issues (Finn et al., 2017; Vanderwal et al., 2015, 2017). Subjects have been noted to exhibit less head movement if they are completing a task versus if they are resting (Huijbers et al., 2017; Vanderwal et al., 2015). Given the effects of motion previously discussed, it is reasonable that, by decreasing movement, subjects might be more identifiable from scan-to-scan. In addition, completing a task forces subjects to stay awake (Vanderwal et al., 2015), which again might render subjects more identifiable; if a subject is asleep in one resting-state scan and awake in the other, it is likely different patterns of FC are underlying these distinct states, and thus intrasubject correlation will be reduced. Taken together, tasks and naturalistic stimuli seem to augment unique, individual-specific FC patterns, providing higher identifications than relatively noisy rest conditions. In addition, in the context of connectome-based identification, tasks offer many pragmatic advantages that rest scans do not.

IDENTIFICATION RELEVANCE TO BEHAVIOR

Although connectome-based identification is a powerful technique to highlight and study individual differences in FC, and it was perhaps initially unexpected that the individual connectome was unique to this extent, one would never use functional connectome mapping simply for this purpose. The power of this approach lies in understanding interindividual differences in FC patterns, allowing researchers to leverage this variability to produce meaningful models related to behavior or cognition. To this end, we have developed a technique, called connectome-based predictive modeling (CPM; Shen et al., 2017), that uses FC patterns to predict levels of behavioral or cognitive variables (Fig. 3). A full discussion of CPM is outside the scope of this chapter, though we direct the reader to Shen et al. (2017) for a discussion of the method and practical suggestions for implementing it into a research pipeline. Utilizing only FC data, the method has been used to build informative models predicting fluid intelligence (gF; Finn et al., 2015) and sustained attention (Rosenberg et al., 2016a). In both of these studies, FC models could predict the score of previously unseen individuals, and in the case of the attention study, the models built utilizing the FC patterns of healthy adults could generalize to predict attention-deficit/hyperactivity disorder (ADHD) symptom scores in an independent dataset of children and adolescents with ADHD diagnoses. Other studies using different methods have also used FC patterns to predict gF (Hearne et al., 2016), analyze different subtypes of depression and their subsequent response to treatment (Drysdale et al., 2017), and psychiatric symptom scores in children and adolescents (Kaufmann et al., 2017). Thus, these results reinforce that, although much work remains to be conducted, individual differences in FC patterns allow meaningful predictions about behavioral, cognitive, and disease-specific features.

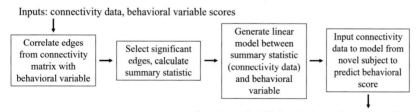

FIG. 3

An overview of the CPM pipeline. Connectivity data are correlated with a behavioral measure to identity significant edges. These edges are summarized to a single value, and a model is built relating the summary statistic and the behavioral variable. Connectivity data from a novel subject is input into the model, and a predicted behavioral score is generated.

CONCLUSIONS AND OUTLOOK

In conclusion, the individual connectome is unique, and methods such as connectome-based predictive modeling are emerging that allow the link between FC patterns and behavior to be explored. Functional identification procedures provide a means for characterizing individual variability in brain function. Here, we have detailed how to conduct the procedure using FC data and highlighted results in the application of this approach to different populations. Common anatomical regions that contribute to high ID rates and thus individual uniqueness primarily include frontal and parietal cortical regions. The experimental factors that can influence identification accuracy and the effect of scan acquisition protocols were discussed. Naturalistic viewing and/or task-based acquisitions in general show better identification rates highlighting how brain state can influence individual variability. Finally, and most importantly, emerging evidence suggests that the individual functional connectome contains a wealth of information related to behavior and possibly symptoms in clinical populations, and we are in the early stages of building models to extract this information. Despite what has been learned about the individual connectome, a number of questions persist (Box 1). Although answers await further study, the investigation of individual subjects in neuroimaging should continue to yield important discoveries shedding light on fundamental aspects of brain function in health and disease.

BOX 1 OPEN QUESTIONS REGARDING CONNECTOME-BASED IDENTIFICATION STUDIES

- At what point in development does a unique and stable functional connectome emerge? Under what conditions do you no longer look the same?
- Do unique edges helpful to identification remain so during development and across the life span?
- How does the connectome vary over the course of short time frames (days, hours, minutes), and what do these variations mean?
- How does the FC profile vary in different disease states? What specific aspects of the connectome can be used to predict disease onset, progression, and response to treatments?
- Are there consistent patterns when a subject is misidentified (i.e., do females tend to be misidentified as females, a patient with depression misidentified as a depressed individual, etc.)?
- Besides anatomical location, are there consistent patterns in highly unique edges (i.e., do long range edges contribute to identification more than short-range edges, or do edges on the lateral aspects/periphery of the brain contribute more than midline edges, are there consistent hemispheric differences in contribution, or are inter- versus intrahemispheric edges more informative)? How do these factors change with development and across the life span?
- How are individual differences captured by FC patterns expressed at the neuronal, synaptic, and subcellular level?
- Are there other factors that have not been accounted for driving the high within-subject correlations and identification rates? If so, what are they and how can they be further exploited?

REFERENCES

Airan, R.D., Vogelstein, J.T., Pillai, J.J., Caffo, B., Pekar, J.J., Sair, H.I., 2016. Factors affecting characterization and localization of interindividual differences in functional connectivity using MRI. Hum. Brain Mapp. 37, 1986–1997.

Barnes, A., Bullmore, E.T., Suckling, J., 2009. Endogenous human brain dynamics recover slowly following cognitive effort. PLoS One 4, e6626.

Birn, R.M., Molloy, E.K., Patriat, R., Parker, T., Meier, T.B., Kirk, G.R., Nair, V.A., Meyerand, M.E., Prabhakaran, V., 2013. The effect of scan length on the reliability of resting-state fMRI connectivity estimates. Neuroimage 83, 550–558.

Biswal, B.B., Mennes, M., Zuo, X.N., Gohel, S., Kelly, C., Smith, S.M., Beckmann, C.F., Adelstein, J.S., Buckner, R.L., Colcombe, S., Dogonowski, A.M., Ernst, M., Fair, D., Hampson, M., Hoptman, M.J., Hyde, J.S., Kiviniemi, V.J., Kotter, R., Li, S.J., Lin, C.P., Lowe, M.J., Mackay, C., Madden, D.J., Madsen, K.H., Margulies, D.S., Mayberg, H.S., McMahon, K., Monk, C.S., Mostofsky, S.H., Nagel, B.J., Pekar, J.J., Peltier, S.J., Petersen, S.E., Riedl, V., Rombouts, S.A., Rypma, B., Schlaggar, B.L., Schmidt, S., Seidler, R.D., Siegle, G.J., Sorg, C., Teng, G.J., Veijola, J., Villringer, A., Walter, M., Wang, L., Weng, X.C., Whitfield-Gabrieli, S., Williamson, P., Windischberger, C., Zang, Y.F., Zhang, H.Y., Castellanos, F.X., Milham, M.P., 2010. Toward discovery science of human brain function. Proc. Natl. Acad. Sci. U. S. A. 107, 4734–4739.

Blumberg, H.P., Leung, H.C., Skudlarski, P., Lacadie, C.M., Fredericks, C.A., Harris, B.C., Charney, D.S., Gore, J.C., Krystal, J.H., Peterson, B.S., 2003. A functional magnetic resonance imaging study of bipolar disorder: state- and trait-related dysfunction in ventral prefrontal cortices. Arch. Gen. Psychiatry 60, 601–609.

Brenhouse, H.C., Andersen, S.L., 2011. Developmental trajectories during adolescence in males and females: a cross-species understanding of underlying brain changes. Neurosci. Biobehav. Rev. 35, 1687–1703.

Buckner, R.L., Krienen, F.M., Castellanos, A., Diaz, J.C., Yeo, B.T., 2011. The organization of the human cerebellum estimated by intrinsic functional connectivity. J. Neurophysiol. 106, 2322–2345.

Burnett, S., Sebastian, C., Cohen Kadosh, K., Blakemore, S.J., 2011. The social brain in adolescence: evidence from functional magnetic resonance imaging and behavioural studies. Neurosci. Biobehav. Rev. 35, 1654–1664.

Casey, B.J., Jones, R.M., 2010. Neurobiology of the adolescent brain and behavior: implications for substance use disorders. J. Am. Acad. Child Adolesc. Psychiatry 49, 1189–1201. quiz 1285.

Casey, B.J., Giedd, J.N., Thomas, K.M., 2000. Structural and functional brain development and its relation to cognitive development. Biol. Psychol. 54, 241–257.

Casey, B.J., Jones, R.M., Hare, T.A., 2008. The adolescent brain. Ann. N. Y. Acad. Sci. 1124, 111–126.

Casey, B.J., Jones, R.M., Levita, L., Libby, V., Pattwell, S.S., Ruberry, E.J., Soliman, F., Somerville, L.H., 2010. The storm and stress of adolescence: insights from human imaging and mouse genetics. Dev. Psychobiol. 52, 225–235.

Chang, C., Leopold, D.A., Scholvinck, M.L., Mandelkow, H., Picchioni, D., Liu, X., Ye, F.Q., Turchi, J.N., Duyn, J.H., 2016. Tracking brain arousal fluctuations with fMRI. Proc. Natl. Acad. Sci. U. S. A. 113, 4518–4523.

Christoff, K., Gordon, A.M., Smallwood, J., Smith, R., Schooler, J.W., 2009. Experience sampling during fMRI reveals default network and executive system contributions to mind wandering. Proc. Natl. Acad. Sci. U. S. A. 106, 8719–8724.

Collins, F.S., Brooks, L.D., Chakravarti, A., 1998. A DNA polymorphism discovery resource for research on human genetic variation. Genome Res. 8, 1229–1231.

Craddock, R.C., James, G.A., Holtzheimer 3rd, P.E., Hu, X.P., Mayberg, H.S., 2012. A whole brain fMRI atlas generated via spatially constrained spectral clustering. Hum. Brain Mapp. 33, 1914–1928.

Donegan, N.H., Sanislow, C.A., Blumberg, H.P., Fulbright, R.K., Lacadie, C., Skudlarski, P., Gore, J.C., Olson, I.R., McGlashan, T.H., Wexler, B.E., 2003. Amygdala hyperreactivity in borderline personality disorder: implications for emotional dysregulation. Biol. Psychiatry 54, 1284–1293.

Drysdale, A.T., Grosenick, L., Downar, J., Dunlop, K., Mansouri, F., Meng, Y., Fetcho, R.N., Zebley, B., Oathes, D.J., Etkin, A., Schatzberg, A.F., Sudheimer, K., Keller, J., Mayberg, H.S., Gunning, F.M., Alexopoulos, G.S., Fox, M.D., Pascual-Leone, A., Voss, H.U., Casey, B.J., Dubin, M.J., Liston, C., 2017. Resting-state connectivity biomarkers define neurophysiological subtypes of depression. Nat. Med. 23, 28–38.

Duncan, N.W., Northoff, G., 2013. Overview of potential procedural and participant-related confounds for neuroimaging of the resting state. J. Psychiatry Neurosci. 38, 84–96.

Durston, S., Tottenham, N.T., Thomas, K.M., Davidson, M.C., Eigsti, I.M., Yang, Y., Ulug, A.M., Casey, B.J., 2003. Differential patterns of striatal activation in young children with and without ADHD. Biol. Psychiatry 53, 871–878.

Finn, E.S., Todd Constable, R., 2016. Individual variation in functional brain connectivity: implications for personalized approaches to psychiatric disease. Dialogues Clin. Neurosci. 18, 277–287.

Finn, E.S., Shen, X., Scheinost, D., Rosenberg, M.D., Huang, J., Chun, M.M., Papademetris, X., Constable, R.T., 2015. Functional connectome fingerprinting: identifying individuals using patterns of brain connectivity. Nat. Neurosci. 18, 1664–1671.

Finn, E.S., Scheinost, D., Finn, D.M., Shen, X., Papademetris, X., Constable, R.T., 2017. Can brain state be manipulated to emphasize individual differences in functional connectivity. Neuroimage.

Fischl, B., van der Kouwe, A., Destrieux, C., Halgren, E., Segonne, F., Salat, D.H., Busa, E., Seidman, L.J., Goldstein, J., Kennedy, D., Caviness, V., Makris, N., Rosen, B., Dale, A.M., 2004. Automatically parcellating the human cerebral cortex. Cereb. Cortex 14, 11–22.

Frazer, K.A., Murray, S.S., Schork, N.J., Topol, E.J., 2009. Human genetic variation and its contribution to complex traits. Nat. Rev. Genet. 10, 241–251.

Gabrieli, J.D., Ghosh, S.S., Whitfield-Gabrieli, S., 2015. Prediction as a humanitarian and pragmatic contribution from human cognitive neuroscience. Neuron 85, 11–26.

Gee, D.G., Gabard-Durnam, L.J., Flannery, J., Goff, B., Humphreys, K.L., Telzer, E.H., Hare, T.A., Bookheimer, S.Y., Tottenham, N., 2013. Early developmental emergence of human amygdala-prefrontal connectivity after maternal deprivation. Proc. Natl. Acad. Sci. U. S. A. 110, 15638–15643.

Gee, D.G., Gabard-Durnam, L., Telzer, E.H., Humphreys, K.L., Goff, B., Shapiro, M., Flannery, J., Lumian, D.S., Fareri, D.S., Caldera, C., Tottenham, N., 2014. Maternal buffering of human amygdala-prefrontal circuitry during childhood but not during adolescence. Psychol. Sci. 25, 2067–2078.

Giedd, J.N., Rapoport, J.L., 2010. Structural MRI of pediatric brain development: what have we learned and where are we going? Neuron 67, 728–734.

Giedd, J.N., Lalonde, F.M., Celano, M.J., White, S.L., Wallace, G.L., Lee, N.R., Lenroot, R.K., 2009. Anatomical brain magnetic resonance imaging of typically developing children and adolescents. J. Am. Acad. Child Adolesc. Psychiatry 48, 465–470.

Glasser, M.F., Coalson, T.S., Robinson, E.C., Hacker, C.D., Harwell, J., Yacoub, E., Ugurbil, K., Andersson, J., Beckmann, C.F., Jenkinson, M., Smith, S.M., Van Essen, D.C., 2016. A multi-modal parcellation of human cerebral cortex. Nature 536, 171–178.

Gogtay, N., Giedd, J.N., Lusk, L., Hayashi, K.M., Greenstein, D., Vaituzis, A.C., Nugent III, T.F., Herman, D.H., Clasen, L.S., Toga, A.W., Rapoport, J.L., Thompson, P.M., 2004. Dynamic mapping of human cortical development during childhood through early adulthood. Proc. Natl. Acad. Sci. U. S. A. 101, 8174–8179.

Gordon, E.M., Laumann, T.O., Adeyemo, B., Huckins, J.F., Kelley, W.M., Petersen, S.E., 2016. Generation and evaluation of a cortical area parcellation from resting-state correlations. Cereb. Cortex 26, 288–303.

Gorgolewski, K.J., Lurie, D., Urchs, S., Kipping, J.A., Craddock, R.C., Milham, M.P., Margulies, D.S., Smallwood, J., 2014. A correspondence between individual differences in the brain's intrinsic functional architecture and the content and form of self-generated thoughts. PLoS One 9, e97176.

Greicius, M., 2008. Resting-state functional connectivity in neuropsychiatric disorders. Curr. Opin. Neurol. 21, 424–430.

Hannon, E., Lunnon, K., Schalkwyk, L., Mill, J., 2015. Interindividual methylomic variation across blood, cortex, and cerebellum: implications for epigenetic studies of neurological and neuropsychiatric phenotypes. Epigenetics 10, 1024–1032.

Hare, T.A., Tottenham, N., Galvan, A., Voss, H.U., Glover, G.H., Casey, B.J., 2008. Biological substrates of emotional reactivity and regulation in adolescence during an emotional go-nogo task. Biol. Psychiatry 63, 927–934.

Hariri, A.R., Mattay, V.S., Tessitore, A., Kolachana, B., Fera, F., Goldman, D., Egan, M.F., Weinberger, D.R., 2002. Serotonin transporter genetic variation and the response of the human amygdala. Science 297, 400–403.

Harrison, B.J., Pujol, J., Ortiz, H., Fornito, A., Pantelis, C., Yucel, M., 2008. Modulation of brain resting-state networks by sad mood induction. PLoS One 3, e1794.

Hearne, L.J., Mattingley, J.B., Cocchi, L., 2016. Functional brain networks related to individual differences in human intelligence at rest. Sci. Rep. 6, 32328.

Heils, A., Teufel, A., Petri, S., Stober, G., Riederer, P., Bengel, D., Lesch, K.P., 1996. Allelic variation of human serotonin transporter gene expression. J. Neurochem. 66, 2621–2624.

Hines, M., 2010. Sex-related variation in human behavior and the brain. Trends Cogn. Sci. 14, 448–456.

Horien, C., Noble, S., Finn, E.S., Shen, X., Scheinost, D., Constable, R.T., 2017a. Considering factors affecting the connectome-based identification process: comment on Waller et al. Neuroimage 169, 172–175.

Horien, C., Shen, S., Scheinost, D., Constable, R.T., 2017b. Individual connectomes are unique and stable in the developing brain from adolescence to young adulthood. bioRxiv.

Houston, I., Peter, C.J., Mitchell, A., Straubhaar, J., Rogaev, E., Akbarian, S., 2013. Epigenetics in the human brain. Neuropsychopharmacology 38, 183–197.

Huijbers, W., Van Dijk, K.R., Boenniger, M.M., Stirnberg, R., Breteler, M.M., 2017. Less head motion during MRI under task than resting-state conditions. Neuroimage 147, 111–120.

Iafrate, A.J., Feuk, L., Rivera, M.N., Listewnik, M.L., Donahoe, P.K., Qi, Y., Scherer, S.W., Lee, C., 2004. Detection of large-scale variation in the human genome. Nat. Genet. 36, 949–951.

Illingworth, R.S., Gruenewald-Schneider, U., De Sousa, D., Webb, S., Merusi, C., Kerr, A.R., James, K.D., Smith, C., Walker, R., Andrews, R., Bird, A.P., 2015. Inter-individual variability contrasts with regional homogeneity in the human brain DNA methylome. Nucleic Acids Res. 43, 732–744.

Johnson, S.B., Blum, R.W., Giedd, J.N., 2009. Adolescent maturity and the brain: the promise and pitfalls of neuroscience research in adolescent health policy. J. Adolesc. Health 45, 216–221.

Kaufmann, T., Alnaes, D., Doan, N.T., Brandt, C.L., Andreassen, O.A., Westlye, L.T., 2017. Delayed stabilization and individualization in connectome development are related to psychiatric disorders. Nat. Neurosci. 20, 513–515.

Kilpatrick, L.A., Zald, D.H., Pardo, J.V., Cahill, L.F., 2006. Sex-related differences in amygdala functional connectivity during resting conditions. Neuroimage 30, 452–461.

Laumann, T.O., Gordon, E.M., Adeyemo, B., Snyder, A.Z., Joo, S.J., Chen, M.Y., Gilmore, A.W., McDermott, K.B., Nelson, S.M., Dosenbach, N.U., Schlaggar, B.L., Mumford, J.A., Poldrack, R.A., Petersen, S.E., 2015. Functional system and areal organization of a highly sampled individual human brain. Neuron 87, 657–670.

Li, C.S., Milivojevic, V., Constable, R.T., Sinha, R., 2005. Recent cannabis abuse decreased stress-induced BOLD signals in the frontal and cingulate cortices of cocaine dependent individuals. Psychiatry Res. 140, 271–280.

Liu, W., Wei, D., Chen, Q., Yang, W., Meng, J., Wu, G., Bi, T., Zhang, Q., Zuo, X.N., Qiu, J., 2017. Longitudinal test-retest neuroimaging data from healthy young adults in southwest China. Sci Data 4, 170017.

Mueller, S., Wang, D., Fox, M.D., Yeo, B.T., Sepulcre, J., Sabuncu, M.R., Shafee, R., Lu, J., Liu, H., 2013. Individual variability in functional connectivity architecture of the human brain. Neuron 77, 586–595.

Mueller, S., Wang, D., Fox, M.D., Pan, R., Lu, J., Li, K., Sun, W., Buckner, R.L., Liu, H., 2015. Reliability correction for functional connectivity: Theory and implementation. Hum. Brain Mapp. 36, 4664–4680.

Nettle, D., 2006. The evolution of personality variation in humans and other animals. Am. Psychol. 61, 622–631.

Noble, S., Spann, M.N., Tokoglu, F., Shen, X., Constable, R.T., Scheinost, D., 2017. Influences on the test-retest reliability of functional connectivity mri and its relationship with behavioral utility. Cereb. Cortex 1–15.

Pannunzi, M., Hindriks, R., Bettinardi, R.G., Wenger, E., Lisofsky, N., Martensson, J., Butler, O., Filevich, E., Becker, M., Lochstet, M., Kuhn, S., Deco, G., 2017. Resting-state fMRI correlations: from link-wise unreliability to whole brain stability. Neuroimage 157, 250–262.

Patriat, R., Molloy, E.K., Meier, T.B., Kirk, G.R., Nair, V.A., Meyerand, M.E., Prabhakaran, V., Birn, R.M., 2013. The effect of resting condition on resting-state fMRI reliability and consistency: a comparison between resting with eyes open, closed, and fixated. Neuroimage 78, 463–473.

Poole, V.N., Robinson, M.E., Singleton, O., DeGutis, J., Milberg, W.P., McGlinchey, R.E., Salat, D.H., Esterman, M., 2016. Intrinsic functional connectivity predicts individual differences in distractibility. Neuropsychologia 86, 176–182.

Potenza, M.N., Leung, H.C., Blumberg, H.P., Peterson, B.S., Fulbright, R.K., Lacadie, C.M., Skudlarski, P., Gore, J.C., 2003. An FMRI Stroop task study of ventromedial prefrontal cortical function in pathological gamblers. Am. J. Psychiatry 160, 1990–1994.

Power, J.D., Schlaggar, B.L., Petersen, S.E., 2015. Recent progress and outstanding issues in motion correction in resting state fMRI. Neuroimage 105, 536–551.

Rahim, N.G., Harismendy, O., Topol, E.J., Frazer, K.A., 2008. Genetic determinants of phenotypic diversity in humans. Genome Biol. 9, 215.

Redon, R., Ishikawa, S., Fitch, K.R., Feuk, L., Perry, G.H., Andrews, T.D., Fiegler, H., Shapero, M.H., Carson, A.R., Chen, W., Cho, E.K., Dallaire, S., Freeman, J.L., Gonzalez, J.R., Gratacos, M., Huang, J., Kalaitzopoulos, D., Komura, D., MacDonald, J.R., Marshall, C.R., Mei, R., Montgomery, L., Nishimura, K., Okamura, K., Shen, F., Somerville, M.J., Tchinda, J., Valsesia, A., Woodwark, C., Yang, F., Zhang, J., Zerjal, T., Zhang, J., Armengol, L., Conrad, D.F., Estivill, X., Tyler-Smith, C., Carter, N.P., Aburatani, H., Lee, C., Jones, K.W., Scherer, S.W., Hurles, M.E., 2006. Global variation in copy number in the human genome. Nature 444, 444–454.

Rosenberg, M.D., Finn, E.S., Scheinost, D., Papademetris, X., Shen, X., Constable, R.T., Chun, M.M., 2016a. A neuromarker of sustained attention from whole-brain functional connectivity. Nat. Neurosci. 19, 165–171.

Rosenberg, M.D., Zhang, S., Hsu, W.T., Scheinost, D., Finn, E.S., Shen, X., Constable, R.T., Li, C.S., Chun, M.M., 2016b. Methylphenidate modulates functional network connectivity to enhance attention. J. Neurosci. 36, 9547–9557.

Rosenberg, M.D., Hsu, W.T., Scheinost, D., Constable, R.T., Chun, M.M., 2018. Connectome-based models predict separable components of attention in novel individuals. J. Cogn. Neurosci. 30, 160–173.

Satterthwaite, T.D., Wolf, D.H., Loughead, J., Ruparel, K., Elliott, M.A., Hakonarson, H., Gur, R.C., Gur, R.E., 2012. Impact of in-scanner head motion on multiple measures of functional connectivity: relevance for studies of neurodevelopment in youth. Neuroimage 60, 623–632.

Satterthwaite, T.D., Wolf, D.H., Roalf, D.R., Ruparel, K., Erus, G., Vandekar, S., Gennatas, E.D., Elliott, M.A., Smith, A., Hakonarson, H., Verma, R., Davatzikos, C., Gur, R.E., Gur, R.C., 2015. Linked sex differences in cognition and functional connectivity in youth. Cereb. Cortex 25, 2383–2394.

Satterthwaite, T.D., Connolly, J.J., Ruparel, K., Calkins, M.E., Jackson, C., Elliott, M.A., Roalf, D.R., Ryan Hopsona, K.P., Behr, M., Qiu, H., Mentch, F.D., Chiavacci, R., Sleiman, P.M., Gur, R.C., Hakonarson, H., Gur, R.E., 2016. The Philadelphia Neurodevelopmental Cohort: a publicly available resource for the study of normal and abnormal brain development in youth. Neuroimage 124, 1115–1119.

Schaefer, A., Kong, R., Gordon, E.M., Laumann, T.O., Zuo, X.N., Holmes, A.J., Eickhoff, S.B., Yeo, B.T.T., 2017. Local-global parcellation of the human cerebral cortex from intrinsic functional connectivity MRI. Cereb. Cortex 1–20. https://doi.org/10.1093/cercor/bhx179.

Scheinost, D., Finn, E.S., Tokoglu, F., Shen, X., Papademetris, X., Hampson, M., Constable, R.T., 2015. Sex differences in normal age trajectories of functional brain networks. Hum. Brain Mapp. 36, 1524–1535.

Shah, L.M., Cramer, J.A., Ferguson, M.A., Birn, R.M., Anderson, J.S., 2016. Reliability and reproducibility of individual differences in functional connectivity acquired during task and resting state. Brain Behav. 6, e00456.

Shaywitz, B.A., Shaywitz, S.E., Pugh, K.R., Mencl, W.E., Fulbright, R.K., Skudlarski, P., Constable, R.T., Marchione, K.E., Fletcher, J.M., Lyon, G.R., Gore, J.C., 2002. Disruption of posterior brain systems for reading in children with developmental dyslexia. Biol. Psychiatry 52, 101–110.

Shen, X., Tokoglu, F., Papademetris, X., Constable, R.T., 2013. Groupwise whole-brain parcellation from resting-state fMRI data for network node identification. Neuroimage 82, 403–415.

Shen, X., Finn, E.S., Scheinost, D., Rosenberg, M.D., Chun, M.M., Papademetris, X., Constable, R.T., 2017. Using connectome-based predictive modeling to predict individual behavior from brain connectivity. Nat. Protoc. 12, 506–518.

Shirer, W.R., Ryali, S., Rykhlevskaia, E., Menon, V., Greicius, M.D., 2012. Decoding subject-driven cognitive states with whole-brain connectivity patterns. Cereb. Cortex 22, 158–165.

Smith, S.M., Beckmann, C.F., Andersson, J., Auerbach, E.J., Bijsterbosch, J., Douaud, G., Duff, E., Feinberg, D.A., Griffanti, L., Harms, M.P., Kelly, M., Laumann, T., Miller, K.L., Moeller, S., Petersen, S., Power, J., Salimi-Khorshidi, G., Snyder, A.Z., Vu, A.T., Woolrich, M.W., Xu, J., Yacoub, E., Ugurbil, K., Van Essen, D.C., Glasser, M.F., Consortium, W.U.-M.H., 2013. Resting-state fMRI in the Human Connectome Project. Neuroimage 80, 144–168.

Sweatt, J.D., Tamminga, C.A., 2016. An epigenomics approach to individual differences and its translation to neuropsychiatric conditions. Dialogues Clin. Neurosci. 18, 289–298.

Tagliazucchi, E., Laufs, H., 2014. Decoding wakefulness levels from typical fMRI resting-state data reveals reliable drifts between wakefulness and sleep. Neuron 82, 695–708.

Tamnes, C.K., Walhovd, K.B., Dale, A.M., Ostby, Y., Grydeland, H., Richardson, G., Westlye, L.T., Roddey, J.C., Hagler Jr., D.J., Due-Tonnessen, P., Holland, D., Fjell, A.M., Alzheimer's Disease Neuroimaging Initiative, 2013. Brain development and aging: overlapping and unique patterns of change. Neuroimage 68, 63–74.

Tau, G.Z., Peterson, B.S., 2010. Normal development of brain circuits. Neuropsycho-pharmacology 35, 147–168.

Tavor, I., Parker Jones, O., Mars, R.B., Smith, S.M., Behrens, T.E., Jbabdi, S., 2016. Task-free MRI predicts individual differences in brain activity during task performance. Science 352, 216–220.

Tung, K.C., Uh, J., Mao, D., Xu, F., Xiao, G., Lu, H., 2013. Alterations in resting functional connectivity due to recent motor task. Neuroimage 78, 316–324.

Van Essen, D.C., Smith, S.M., Barch, D.M., Behrens, T.E., Yacoub, E., Ugurbil, K., Consortium, W.U.-M.H., 2013. The WU-minn human connectome project: an overview. Neuroimage 80, 62–79.

Vanderwal, T., Kelly, C., Eilbott, J., Mayes, L.C., Castellanos, F.X., 2015. Inscapes: a movie paradigm to improve compliance in functional magnetic resonance imaging. Neuroimage 122, 222–232.

Vanderwal, T., Eilbott, J., Finn, E.S., Craddock, R.C., Turnbull, A., Castellanos, F.X., 2017. Individual differences in functional connectivity during naturalistic viewing conditions. Neuroimage 157, 521–530.

Waller, L., Walter, H., Kruschwitz, J.D., Reuter, L., Muller, S., Erk, S., Veer, I.M., 2017. Evaluating the replicability, specificity, and generalizability of connectome fingerprints. Neuroimage 158, 371–377.

Yeo, B.T., Krienen, F.M., Sepulcre, J., Sabuncu, M.R., Lashkari, D., Hollinshead, M., Roffman, J.L., Smoller, J.W., Zollei, L., Polimeni, J.R., Fischl, B., Liu, H., Buckner, R.L., 2011. The organization of the human cerebral cortex estimated by intrinsic functional connectivity. J. Neurophysiol. 106, 1125–1165.

Zilles, K., Armstrong, E., Schleicher, A., Kretschmann, H.J., 1988. The human pattern of gyri-fication in the cerebral cortex. Anat. Embryol. 179, 173–179.

Zuo, X.N., Anderson, J.S., Bellec, P., Birn, R.M., Biswal, B.B., Blautzik, J., Breitner, J.C., Buckner, R.L., Calhoun, V.D., Castellanos, F.X., Chen, A., Chen, B., Chen, J., Chen, X., Colcombe, S.J., Courtney, W., Craddock, R.C., Di Martino, A., Dong, H.M., Fu, X., Gong, Q., Gorgolewski, K.J., Han, Y., He, Y., He, Y., Ho, E., Holmes, A., Hou, X.H., Huckins, J., Jiang, T., Jiang, Y., Kelley, W., Kelly, C., King, M., LaConte, S.M., Lainhart, J.E., Lei, X., Li, H.J., Li, K., Li, K., Lin, Q., Liu, D., Liu, J., Liu, X., Liu, Y., Lu, G., Lu, J., Luna, B., Luo, J., Lurie, D., Mao, Y., Margulies, D.S., Mayer, A.R., Meindl, T., Meyerand, M.E., Nan, W., Nielsen, J.A., O'Connor, D., Paulsen, D., Prabhakaran, V., Qi, Z., Qiu, J., Shao, C., Shehzad, Z., Tang, W., Villringer, A., Wang, H., Wang, K., Wei, D., Wei, G.X., Weng, X.C., Wu, X., Xu, T., Yang, N., Yang, Z., Zang, Y.F., Zhang, L., Zhang, Q., Zhang, Z., Zhang, Z., Zhao, K., Zhen, Z., Zhou, Y., Zhu, X.T., Milham, M.P., 2014. An open science resource for establishing reliability and reproducibility in functional connectomics. Sci Data 1, 140049.

Dysfunctional brain network organization in neurodevelopmental disorders

5

Teague R. Henry, Jessica R. Cohen

Department of Psychology and Neuroscience, University of North Carolina at Chapel Hill,
Chapel Hill, NC, United States

CHAPTER OUTLINE

The term "neurodevelopmental disorder" covers a wide range of brain disorders that arise from atypical brain development beginning early in life (and in some cases prenatally), and continuing throughout the life span. The etiology of some of these disorders is well known. For example, Fragile X Syndrome is a genetic disorder that arises from a lengthening of the FMR1 gene on the X chromosome, a mutation that leads to disrupted brain development (Hagerman and Hagerman, 2004). For others, such as attention deficit hyperactivity disorder (ADHD) and autism spectrum disorder (ASD), the causes are considerably less clear and are likely much more heterogeneous. Recent research investigating the brain basis of many of these neurodevelopmental disorders has led to the hypothesis that the pathophysiology of these disorders involves dysfunctional brain network organization, rather than disrupted functioning of individual brain regions.

Before delving into the research that has led to such a hypothesis, it is critical to understand how brain network organization is measured in humans. Graph theoretical tools from the field of mathematics can be used to describe the brain as a graph, or a network, composed of nodes and edges. The term *node* refers to specific brain regions of interest (ROIs). These ROIs can be anatomically defined, such as the amygdala, or they can be functionally defined, such as a portion of the inferior

Connectomics. https://doi.org/10.1016/B978-0-12-813838-0.00005-4

frontal gyrus that relates to a specific set of cognitive processes. To place the ROIs within the notation of graph theory, we will interchangeably refer to ROIs as nodes. In a brain graph, the term *edge* refers to the connectivity between ROIs (nodes). The operationalization of these connectivity edges will depend on the neuroimaging modality and analysis conducted. A common way of defining a functional connectivity edge is to calculate the correlation between fluctuations in the BOLD signals across time of two ROIs. Alternatively, one can use the spectral coherence between the BOLD signals. A structural connectivity edge typically corresponds to the number of streamlines, or white matter fibers, between a pair of ROIs. In this chapter, we will use the term edge and connection interchangeably, while specifying the modality (functional or structural) of the connection.

Within the whole-brain network, subnetworks are strongly interconnected sets of ROIs. Subnetwork organization can be determined in a data-driven manner using a community detection algorithm (Sporns and Betzel, 2016). Subnetworks can alternately be defined a priori based on existing literature, such as the default mode network (DMN) or the salience network (SN) (Seeley et al., 2007). In human neuroimaging literature, specific subnetworks are often referred to as "networks" due to the history of probing individual subnetworks outside the context of the whole-brain network. Throughout this chapter, we utilize names commonly used in human neuroimaging literature (i.e., DMN), but clarify that they are, in fact, subnetworks within the larger whole-brain network.

One advantage of describing the brain in terms of its network properties is that summary measures of topological organization can be quantified, and therefore single numbers can describe certain network attributes. For example, the average degree of a network is the average, across all nodes, of the number of edges of each node. By utilizing these summary measures, it is possible to describe and analyze large complex networks in simple and easily interpretable ways. For example, the average degree of a network describes how interconnected network nodes are with each other. A framework of network organization emerging as critical for understanding brain function is the balance of integration and segregation across individual subnetworks within the larger whole-brain network (Deco and Kringelbach, 2014; Deco et al., 2015; Shine and Poldrack, 2017; Sporns and Betzel, 2016). Fig. 1 and Table 1 summarize our use of graph theory terminology as well as commonly used metrics describing network attributes.

Dysfunction at the subnetwork level can take multiple forms. Compared to healthy control participants, a clinical population might have reduced or increased connectivity within a subnetwork. As an example, reduced functional connectivity within the DMN during rest has been observed in youth with ADHD (Fair et al., 2010). Alternatively, subnetworks might be aberrantly connected to other subnetworks. For example, increased functional connectivity during rest between a subcortical subnetwork consisting of the basal ganglia and the thalamus, and several cortical subnetworks consisting of primary sensory regions, has been observed in male children with ASD compared with healthy control participants (Cerliani et al., 2015). A growing body of research indicates that there are reliable differences in subnetwork organization between healthy individuals and individuals with

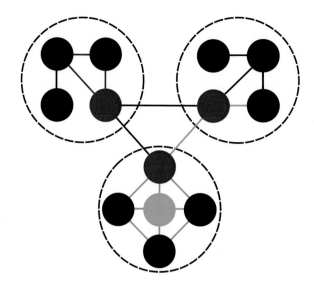

FIG. 1

Network diagram depicting various topological network attributes. *Circles* denote nodes, *lines* denote edges, and the *larger dashed circles* denote individual subnetworks (or communities). See Table 1 for a description of the colored nodes and edges.

Table 1 Topological attributes commonly used to describe brain network organization (see Fig. 1)

Statistic	Description	Figure description/source
Average degree	The average number of edges of a node	The green node has a degree of 4. Freeman (1978)
Clustering coefficient	A proportion describing how interconnected groups of neighboring nodes are to each other	The red edges form a clique of highly interconnected neighboring nodes. Holland and Leinhardt (1971).
Path length	The length of the shortest path between two nodes	The green edges indicate the shortest path between two nodes. Latora and Marchiori (2001).
Nodal efficiency	The average inverse path length from a target node to every other node	NA. Latora and Marchiori (2001).
Global efficiency	The average inverse path length between every pair of nodes	NA. Latora and Marchiori (2001).
Local efficiency	The average nodal efficiency of a node's neighbors	The blue edges represent a highly locally efficient network. Latora and Marchiori (2001).
Modularity	The degree to which a network is segregated into tightly clustered communities	The dashed circles represent individual communities that are highly interconnected. Newman (2006).
Rich club coefficient	The extent to which high degree nodes connect to one another	The red nodes form a rich club. van den Heuvel and Sporns (2011).

psychiatric/neurologic disorders, and a greater understanding of network-based dysfunction may lead to increased knowledge of the mechanisms underlying these disorders (for reviews, see Deco and Kringelbach, 2014; Fornito et al., 2015).

Previous work probing both functional and structural connectivity in neurodevelopmental disorders has primarily examined dysfunction in a hyper/hypoconnectivity framework, where functional and structural connections between regions and between distinct subnetworks are characterized as either stronger or weaker than in healthy individuals. However, the hyper/hypoconnectivity framework does not take into account overall network topology, which has been demonstrated to be an important feature of brain connectivity that differentiates populations (Di Martino et al., 2014; Fornito et al., 2015; Konrad and Eickhoff, 2010). A more recent conceptual framework for understanding functional and structural connectivity patterns emphasizes the importance of balancing the opposing forces of network integration and network segregation (Shine and Poldrack, 2017; Cohen and D'Esposito, 2016; Deco et al., 2015; Sporns, 2013). Network segregation can be conceptualized as strong within-subnetwork connectivity with few interactions across subnetworks, whereas network integration involves greater interactions across distinct subnetworks. This characterization of connectivity patterns focuses less on the strength or weakness of individual functional and structural connections, and instead focuses on how subnetworks of interest are embedded within the overall brain network. This way of thinking opens up avenues of investigation in which the brain dysfunction that underlies neurodevelopmental disorders is not limited to functional or structural *strength* but rather can be described as a disruption of overall *topology*. Notably, previous findings that use the hyper/hypoconnectivity framework can often be recast into a functional or structural integration/segregation approach, making the integration/segregation framework an extension and enrichment of the hyper/hypoconnectivity framework, rather than an opposing approach.

An advantage of the integration/segregation approach is that information about the strength of individual connections is not lost (i.e., whether there is hyper- or hypoconnectivity as related to specific connections or subnetworks), but one is additionally able to calculate summary measures that describe both global (integration) and local (segregation) properties of the whole-brain network (Bullmore and Sporns, 2009; Sporns, 2010). This framework of integration/segregation has been used in recent research to examine properties of cognition in healthy young adults (Cohen and D'Esposito, 2016), preclinical Alzheimer's disease (Brier et al., 2014), and dynamic alternations between integration and segregation of resting state functional connectivity subnetworks (Betzel et al., 2016), among many more topics. In this chapter, we focus on reviewing recent work examining disrupted functional and structural connectivity of brain networks in neurodevelopmental disorders. Given the large emphasis of the literature on ADHD and ASD, we focus on those two disorders. We conclude by discussing extensions of the integration/segregation framework that could help extend our understanding of complex neurodevelopmental disorders. We concentrate on graph theoretic approaches, however, there are two other broad classes of methods for brain connectivity analysis that warrant mentioning. The first is seed-based connectivity analyses, in which a certain voxel or set of voxels is chosen as a "seed" from which to compute connectivity to the rest of the brain (usually operationalized as correlation strength). The second is independent component analysis (ICA), which uses a data-driven method to segment the brain into subnetworks that are spatially or temporally independent from each other. Please see Chapter 7 for more details on the use of these alternative methods to study brain connectivity in neurodevelopmental disorders.

ATTENTION DEFICIT HYPERACTIVITY DISORDER

ADHD is the most commonly diagnosed neurodevelopmental disorder. It is thought to affect approximately 9% of school-aged children in the United States (Center for Disease Control, 2018). This disorder often appears in childhood around year 8 and is characterized by an inability to sustain attention and/or excessive impulsive behavior and hyperactivity (American Psychiatric Association, 2013). FMRI activation studies of children with ADHD have consistently shown hypoactivation in dorsolateral prefrontal cortex (DLPFC), anterior cingulate cortex (ACC), supplementary motor area (SMA), temporoparietal cortical regions, and caudate nucleus during the execution of a variety of inhibition and attention tasks, whereas adults with ADHD show hypoactivation in a number of frontal and striatal regions, in addition to regions in the premotor, parietal, and occipital areas during motor and inhibition tasks (for a review see: Cubillo et al., 2012). Research probing altered reward-related functioning in ADHD has additionally shown hypoactivation in adults, children, and adolescents with ADHD in striatal and ventral-striatal regions during reward anticipation tasks (for reviews, see Cubillo et al., 2012; Plichta and Scheres, 2014). These deficits in frontal and striatal regions have been interpreted previously to indicate a disrupted reward circuit composed of the ventral striatum, thalamus, ACC, and orbitofrontal cortex. As part of their dual pathways model, Sonuga-Barke (2005) suggests that disruptions in this reward circuit are one pathway to ADHD symptomatology, with the other pathway involving disruptions in executive control circuitry. For a review of this literature, see Sonuga-Barke (2002, 2005).

In a large metaanalysis, Cortese et al. (2012) examined a body of ADHD activation studies in an effort to uncover *task agnostic* differences in activation. In addition to observing similar hypoactivation in frontal and parietal regions that had been highlighted in previous reviews (e.g., Cubillo et al., 2012; Plichta and Scheres, 2014), they also found hyperactivation in children with ADHD relative to typically developing children in regions of the DMN and the ventral attention network (VAN), as well as the SMA. In adults with ADHD, hyperactivation was similarly observed in DMN regions but additionally observed in the visual cortex and in regions of the dorsal attention network (DAN). Interestingly, the patterns of hyperactivation were found regardless of task, in contrast with previous work that focused on specific tasks. This suggests a general, rather than a task-specific phenomenon. These three lines of activation findings—hypoactivation in frontal, parietal, and striatal regions during inhibition tasks; hypoactivation in ventral striatal regions during reward anticipation tasks; and hyperactivation in DMN, VAN, and DAN regions regardless of task—suggests that ADHD is characterized by distributed alteration in brain function. However, these activation results are limited to regions studied in isolation, and do not take into account the connections across brain regions.

Functional connectivity measurements can shed light on the nature of the hyper/hypoactivation observed in functional activation studies. As an example, the default mode interference hypothesis (Castellanos and Aoki, 2016; Sonuga-Barke and Castellanos, 2007; Weissman et al., 2006) states that attentional lapses may be due to the DMN "intruding" on regions activated by a task, by either reactivating after temporarily deactivating during the task or failing to deactivate at all (Sonuga-Barke and Castellanos, 2007; Weissman et al., 2006). When applied to individuals with ADHD, the default mode interference hypothesis suggests that the DMN is more likely to "intrude" on task-related regions and subnetworks in individuals with ADHD compared with healthy control participants (Sonuga-Barke and Castellanos, 2007). Reframing this hypothesis in terms of functional connectivity between brain regions suggests that individuals with ADHD should show reduced anticorrelations between the DMN and task-related subnetworks (Sonuga-Barke and Castellanos, 2007). Indeed, research has consistently observed a reduced anticorrelation between the DMN and a variety of task-related subnetworks at rest in children, adolescents, and adults with ADHD (for reviews, see Konrad and Eickhoff, 2010; Posner et al., 2014). Furthermore, studies consistently observe hypoconnectivity *within* the DMN in children and adults with ADHD (for a review, see Posner et al., 2014). This reduced anticorrelation between the DMN and task-related subnetworks can, in part, explain the pattern of hypoactivation of frontal regions and hyperactivation of the DMN that was previously described.

In addition to the default mode interference hypotheses of ADHD, there is a body of literature regarding disruptions in cortico-striatal-thalamic-cortical (CSTC) loops in individuals with ADHD. CSTC loops are neural circuits that project from the cortex to the striatum, to the thalamus, and then back to the cortex. These loops are thought to underlie various cognitive processes (Alexander, 1986; Alexander and Crutcher, 1990). There is a long tradition of functional activation research observing dysfunctional activation in brain regions that are a part of CSTC loops in ADHD (for reviews, see Sonuga-Barke, 2002, 2005; Posner et al., 2014). Recently, a small but growing set of evidence is emerging that functional connectivity of CSTC circuits involved in cognitive and limbic processes are disrupted in individuals with ADHD. Specifically, hypoconnectivity between the putamen and ventral striatum, as well as between the ventral striatum and anterior prefrontal cortex, has been observed in children with ADHD. For a review of this body of literature, refer to Posner et al. (2014).

Within the previously referenced body of literature on the default mode interference and CSTC loop disruption hypotheses, the majority of the studies examined functional connectivity from a strictly hyper/hypoconnectivity framework. Few studies to date examine ADHD from a network integration/segregation framework. In one of these studies, Wang et al. (2009) showed that children with ADHD exhibited decreased global efficiency, a measure of integration, and increased local efficiency (a measure of segregation) relative to healthy individuals. Further, children with ADHD had significantly decreased nodal efficiency of the orbitofrontal cortex and of regions of temporal and occipital cortex, indicating increased segregation of these regions, as well as increased nodal efficiency in the inferior frontal gyrus, indicating increased global integration of this region. In a study in adults with ADHD, it was found that individuals with ADHD exhibited increased modularity, clustering coefficient, and local efficiency, all measures of network segregation, relative to control participants (Lin et al., 2014). The authors additionally found that network segregation was most increased in the frontal cortex, occipital cortex, and subcortical regions, whereas the SMA was more integrated in individuals with ADHD. A study probing integration within the DMN implemented a measure of network homogeneity and found that the DMN exhibited reduced network homogeneity in adults with ADHD relative to age-matched control participants. Reduced network homogeneity was localized to the precuneus, suggesting that the precuneus was more segregated from other regions of the DMN in adults with ADHD (Uddin et al., 2008). More recently, Fair et al. (2013) demonstrated that a pattern classifier was able to successfully differentiate children with the inattentive subtype of ADHD, the combined (inattentive and hyperactive/impulsive) subtype of ADHD, and typically developing children on the basis of functional connectivity patterns. Critically, the brain subnetworks that most differentiated children with the inattentive subtype of ADHD from typically developing children were different from the brain subnetworks that most differentiated children with the combined subtype of ADHD from typically developing children, implying different etiologies of the two subtypes. Specifically, DMN connectivity was most able to differentiate children with the combined subtype of ADHD from typically developing children, whereas frontoparietal network (FPN) and cerebellar connectivity were most able to differentiate children with the inattentive subtype from typically developing children (Fair et al., 2013).

Synthesizing activation, hyper/hypoconnectivity, and integration/segregation results together, this body of research suggests that dysfunctional subnetwork organization in ADHD is not limited to dysfunction within specific subnetworks, such as VAN, DAN, or DMN, but additionally includes disruptions of functional connections between the DMN and task-related subnetworks, as well as disruption of functional connections between cortical and subcortical structures. The small set of integration/segregation studies further suggest that, on a whole-brain level, disruption of functional network organization is mainly due to decreased integration across distinct subnetworks in individuals with ADHD compared with healthy controls (Lin et al., 2014; Wang et al., 2009). A review of graph theoretic functional and structural imaging results in individuals with ADHD (Cao et al., 2014) reached much the same conclusion, noting that there appears to be both general global disruption of structural and functional connectivity, as well as specific loci of dysfunction, such as the DMN, that correlate with ADHD-related behavior. This combination of results suggests several avenues of research that take advantage of the power of an integration/segregation approach. The first is to examine the integration (or segregation) of the DMN with task-related subnetworks during cognitive task performance. Under the default mode interference hypothesis, individuals with ADHD should exhibit a less segregated DMN during cognitive task performance relative to healthy individuals. Further, examining DMN connectivity with other subnetworks in addition to task-related subnetworks would clarify whether the DMN displays reduced segregation specifically with task-related subnetworks or globally reduced segregation with all brain subnetworks. With regard to the overall increase in network segregation, future studies could examine the role of CSTC loops in driving whole-brain integration and segregation in individuals with ADHD. Finally, results of studies probing functional brain network organization can be better informed by synthesizing them with structural brain network analyses. Structural connectivity research in individuals with ADHD has been particularly amenable to the integration/segregation approach in the past, and this literature is what we turn to now.

Recent research probing disrupted structural connectivity in ADHD is consistent with the functional integration/segregation findings discussed previously. Beare et al. (2016), in a study in male children and adolescents with ADHD, used high-angular resolution diffusion imaging (HARDI) and probabilistic tractography to examine differences in structural brain network organization between ADHD and typically developing individuals. They found increased modularity and decreased global efficiency in individuals with ADHD. These findings are consistent with functional network research that has observed increased segregation and decreased integration of distinct brain subnetworks in ADHD. Additionally, they found increased structural connectivity within a subnetwork that encompassed bilateral inferior, middle, and orbitofrontal regions, precentral regions, cingulate cortex, and putamen in individuals with ADHD relative to typically developing individuals (Beare et al., 2016). A study in drug-naïve male children with ADHD similarly observed decreased global efficiency in structural brain network organization compared with typically developing children (Cao et al., 2013). The same study also observed within-network connectivity differences in specific subnetworks in children with ADHD. Specifically, they found decreased structural connectivity within a prefrontal-insular subnetwork and increased structural connectivity within an orbitofrontal-striatal subnetwork in children with ADHD. Combined, literature probing structural and functional network organization in ADHD has consistently found decreased integration and increased segregation in individuals with ADHD (Beare et al., 2016; Cao et al., 2013; Lin et al., 2014; Wang et al., 2009). These results indicate that brain dysfunction in ADHD is described well in terms of a disrupted balance between integration and segregation in brain network organization, both globally (i.e., globally reduced integration) and with regard to specific subnetworks (i.e., specific increased integration of DMN with other subnetworks). Further research exploring disrupted integration and segregation of brain networks in ADHD will increase knowledge of the neural basis of the disorder.

AUTISM SPECTRUM DISORDER

ASD is a complex pervasive neurodevelopmental disorder, the severity of which ranges from very high to very low functioning. The classic symptoms of ASD are marked social deficits, difficulty with language comprehension or production, restricted and repetitive behaviors, and high sensory reactivity (American Psychiatric Association, 2013). ASD is an extremely heterogeneous disorder in terms of presentation. It is also heterogeneous with regard to age of first expression. Typically, the earliest age that a confirmatory diagnosis can be made is 24 months, however in some cases diagnostic symptoms, such as highly repetitive play and lack of social involvement, can present as early as 12 months (Martínez-Pedraza and Carter, 2009).

As with the study of ADHD, neuroimaging studies of ASD began with analyses of fMRI activation in individual brain regions. A recent metaanalysis summarizing a large portion of the brain activation literature suggests two specific patterns of aberrant activation exist in ASD, one related to social processing and the other to nonsocial processing (Di Martino et al., 2009). In the context of social processing, individuals with ASD display decreased activation in the pregenual ACC, anterior rostral medial PFC, amygdala, right anterior insula, and PCC. Each of these regions has been previously implicated in social processing, with the anterior insula and the amygdala of particular interest to ASD research due to the anterior insula's prominent role in the SN (Uddin and Menon, 2009) and the amygdala's connection to facial processing (Baron-Cohen et al., 2000). The metaanalysis also found that, during nonsocial processing, individuals with ASD show decreased activation in dorsal ACC and pre-SMA, regions associated with cognitive control, and increased activation in SMA (Di Martino et al., 2009). Finally, individuals with ASD also demonstrate increased activation in the pregenual ACC during nonsocial processing, similar to what is found during social processing. Based on these results, Di Martino et al. (2009) suggested that disruption of regulatory processes can explain the pattern of hyper- and hypoactivation seen in nonsocial processing. They proposed that the DMN may be the source of this dysfunctional regulation, similar to conclusions from the ADHD connectivity literature.

Functional connectivity research has suggested that there are distributed functional connectivity disruptions throughout the brain in individuals with ASD. Hypoconnectivity has been observed within the DMN, SN (specifically the insular cortex), and the amygdala, while hyperconnectivity has been observed between the striatum and the insular cortex, within the primary motor cortex, and between sensory cortices and the thalamus/basal ganglia (for reviews, see Mueller et al., 2011; Kana et al., 2014; Hull et al., 2016). A recent review summarizing DMN connectivity disruptions in ASD at different ages found that dysfunctional DMN connectivity patterns change across age (Padmanabhan et al., 2017). In children, ASD is associated with hyperconnectivity within the DMN and hypoconnectivity between the DMN and other subnetworks, whereas in adults with ASD, the DMN exhibits hypoconnectivity both within the DMN and between the DMN and other subnetworks. This suggests that, similar to ADHD, the DMN may be a key site of dysfunction in ASD.

In addition to the DMN, another important functional subnetwork with implications for ASD is the SN. This subnetwork, which consists primarily of the insula and the anterior cingulate cortex, is thought to be important for detecting important sources of information across a variety of contexts, as well as switching between distinct task-relevant subnetworks when task demands change (Menon, 2015; Menon and Uddin, 2010; Seeley et al., 2007). Furthermore, in previous work with healthy individuals, the right anterior insula has been shown to play a vital role in the switching of activation between the central executive network (CEN) and the DMN (Menon and Uddin, 2010). Hypoactivation of the anterior insula is implicated in ASD (Uddin and Menon, 2009). Previous work utilizing a machine learning approach to classify subjects into diagnostic categories using functional connectivity showed that connectivity within the SN was better able to distinguish children with ASD from typically developing children than other brain subnetworks such as the DMN (Uddin et al., 2013). In a recent study in primarily male children with ASD, the anterior insula was shown to be hypoconnected to motor, sensory, and visual processing regions in children with ASD compared with typically developing children (Odriozola et al., 2016). Another recent study in children with ASD showed that subjects with ASD had hyperconnectivity within the DMN and right ECN, and hypoconnectivity within the SN and left ECN. Critically, ASD symptomatology was primarily related to dysfunction within the SN (Abbott et al., 2016). Functional connectivity findings with regard to the insula and SN are not entirely consistent, however. Hyperconnectivity has been observed between the insula and the pons in children with ASD (Di Martino et al., 2011), and hyperconnectivity between the SN and sensory cortex has been related to a behavioral measure of sensory overresponsivity in children with ASD (Green et al., 2016). This mix of findings suggests that the SN, and in particular the anterior insula, plays an important role in ASD, but the precise nature of how the SN is embedded within the functional topology of the brain remains unclear.

It is possible that probing functional connectivity in ASD in terms of network integration and network segregation would better describe the dysfunction observed in ASD network organization. Similarly to ADHD, there has been an emerging set of literature that examines ASD from this perspective rather than from a strict hypo/hyperconnectivity perspective. Rudie et al. (2013) found that children with ASD had reduced modularity and increased global efficiency relative to typically developing children, suggesting a more globally integrated functional brain network in ASD. Consistent with these findings, adults with ASD have been shown to display decreased clustering coefficient and decreased characteristic path length, indicating increased integration in ASD (Itahashi et al., 2014). Keown et al. (2017) similarly observed that adolescents and adults with ASD had decreased cohesion (indicating reduced within-subnetwork connectivity or reduced segregation) and increased dispersion (indicating increased across-network integration) relative to healthy control participants. Henry et al. (2017) showed increased whole-brain integration and decreased whole-brain segregation in adolescents and adults with ASD relative to healthy controls and further showed that the somatomotor cortex is a core region driving decreased whole-brain segregation findings. Finally, children with ASD have been shown to have increased functional connectivity within rich club communities, again indicating increased network integration (Ray et al., 2014).

Similarly to research in individuals with ADHD, research of structural connectivity abnormalities in an integration/segregation framework in ASD has begun to appear more frequently. Lewis et al. (2013) used probabilistic tractography to study global network topology of white matter tracts in a study of adult males with ASD. They found a decrease in both local efficiency and global efficiency in individuals with ASD relative to control participants. The authors interpreted this finding as individuals with ASD having more, but weaker, connections overall, which would result in a globally less segregated but also less integrated structural network. Rudie et al. (2013) reported similar findings. They found that, although global efficiency increased with increasing age in typically developing individuals, it decreased with increasing age in individuals with ASD. Conversely, modularity, a measure of network segregation, decreased with increasing age in typically developing individuals, whereas it decreased at a lower rate in individuals with ASD. Consistent with the previously discussed research, Roine et al. (2015) reported reduced global efficiency, increased normalized path length, and decreased strength (the average degree for a weighted network) of the structural network in adult males with ASD. Taken together, this line of research indicates that individuals with ASD have weaker overall white matter network structure, which impacts both network segregation and network integration.

With regard to network segregation, both functional connectivity and structural connectivity literature points to decreased segregation in individuals with ASD. With regard to network integration, however, functional brain networks tend to be more integrated whereas structural brain networks tend to be less integrated. Importantly, in a study that compared both functional and structural network organization in ASD and in healthy individuals, a negative correlation was observed between structural and functional global efficiency (Rudie et al., 2013), confirming previous literature finding increased functional network integration and decreased structural network integration in ASD. This relationship has been observed in other patient populations as well (i.e., in patients with multiple sclerosis; Hawellek et al., 2011).

Taken together, the hyper/hypoconnectivity and integration/segregation findings suggest important further research needs to be conducted to understand dysfunctional brain network organization in individuals with ASD. As an example, understanding how DMN connectivity contributes to overall patterns of whole-brain integration and segregation would shed light on the role of the DMN in dysfunctional network organization in ASD. Probing DMN connectivity in an integration/segregation framework could clarify the findings of hyperconnectivity within the DMN, but hypoconnectivity between the DMN and other subnetworks in children with ASD (Padmanabhan et al., 2017). A similar approach could be applied to examine the role of the SN in brain network organization in ASD. Evidence suggests that ASD is a complex disorder characterized by differences in overall brain topology, and understanding how individual subnetworks within the brain contribute to overall topology would further our understanding of network organization in ASD.

INTEGRATION AND SEGREGATION AS A FRAMEWORK FOR UNDERSTANDING NEURODEVELOPMENTAL DISORDERS: NEXT STEPS

In this chapter, we summarized an emerging framework for understanding disrupted functional and structural connectivity with a focus on balancing network integration and network segregation. We described the general approach this framework invokes, and demonstrated that it complements and extends the more traditional framework of hyper- versus hypoconnectivity. We summarized extant literature probing functional and structural connectivity in both ADHD and ASD, and suggested that the brain basis of both of these disorders is better described as dysfunctional topology rather than altered connectivity strength. These findings suggest several directions that future research can take.

One straightforward direction is to use the integration/segregation framework to localize differences in whole brain integration and segregation to specific subnetworks or ROIs. As was discussed with regard to both ADHD and ASD, several potential lines of research lie in, for example, examining the role of the DMN in overall network structure. Some extant research has done so by examining nodal efficiency (e.g., Wang et al., 2009), but this general approach can be extended to any feature of network organization. Borrowing from network robustness literature in which lesions are simulated, the change in any metric of interest can be examined as a function of the "removal" of nodes and edges (Albert et al., 2000; Callaway et al., 2000). Network robustness methods can assess effects due to the random removal of nodes and edges, as well as the effect of removing specific nodes and edges from a network. This methodology allows researchers to not only describe differences between clinical and control populations in terms of whole-brain network organization, but also in terms of the contribution of specific nodes and edges, as well as subnetworks, to those whole-brain network characteristics. A prominent example of the use of this methodology is found in a study by Achard et al. (2006), in which they demonstrated that the healthy adult brain is more resilient to targeted removal of hub regions than is a scale-free network. Applying this methodology to the study of neurodevelopmental disorders would allow for a more targeted examination of which regions and edges, and therefore subnetworks, are more or less critical to network functioning in disordered populations versus healthy control populations.

A second extension of the integration/segregation framework has already been applied in several studies: the analysis of both structural and functional brain networks within the same subjects. Rudie et al. (2013) demonstrated in a joint analysis of integration metrics from both structural and functional brain networks that global efficiency has an inverse relationship between the two modalities. Ray et al. (2014) performed rich club analysis on both structural and functional data, though they do not implement a joint analysis of the modalities. These analyses highlight one of the strengths of the integration/segregation approach, as it does not require joint analysis of the imaging data but rather joint analysis of the whole-brain summary metrics of network organization derived from constructed connectivity networks. Applying a joint analysis of functional and structural integration/segregation in a whole-brain fashion would allow researchers to examine potential differences in the relationship between structure and function in clinical populations in a fairly easy to interpret fashion. For example, the DMN's contribution to whole-brain functional integration can be examined in tandem with its contribution to whole-brain structural integration, as well as probing differences between patient and control groups. This can be used to examine whether structural/functional coupling differs in different populations. Furthermore, this methodology can be combined with the network robustness methodology examining individual subnetworks and nodes to provide a richer understanding of the nature of structural and functional network organization in specific disorders.

A third extension of the integration/segregation framework would be to characterize the dynamics of functional connectivity (Calhoun et al., 2014). By using dynamic functional connectivity methods to estimate how functional connectivity patterns change throughout the course of a functional brain scan, across-group differences in the dynamics of network integration and segregation can be linked to different patient populations. Given the growing body of evidence that dynamic fluctuations in functional connectivity patterns are related to behavior and cognition in healthy subjects (for reviews, see Cohen, 2017; Kucyi et al., 2018), the combined use of the integration/segregation framework and dynamic functional connectivity methods could help shed light on how dysfunctional connectivity dynamics underlie the differences in cognition and behavior found in patients with neurodevelopmental disorders.

Importantly, given both the similarities and differences in network integration and segregation with regard to ADHD and ASD, future research should directly compare the two populations, both in individuals that have *either* ADHD or ASD, and in individuals with *comorbid* ADHD/ASD (Uddin et al., 2017). For example, one study has included both children with ADHD and children with ASD in an analysis of structural and functional rich club organization. When focusing on structural network organization, a pattern of structural hypoconnectivity within rich club regions and hyperconnectivity outside of the rich club in children with ADHD was found, whereas conversely structural hyperconnectivity within rich-club regions (though with the caveat that these connections were weak) and hypoconnectivity outside of the rich club was found in children with ASD (Ray et al., 2014). Further, when characterizing functional connectivity networks, Ray and colleagues demonstrated that children with ADHD showed significantly higher functional connectivity outside of the rich club than both typically developing children and individuals with ASD, although there were no differences in functional connectivity outside of the rich club between typically developing children and children with ASD. This suggests that this hyperconnectivity outside of the rich club can act as a distinctive neural marker for individuals with ADHD. In a single study, these results support the general findings of increased global integration in ASD and increased segregation in ADHD. When comparing differences in network organization across ADHD and ASD populations, these results further highlight the differences in dysfunctional network organization that underlies each disorder.

In closing, neurodevelopmental disorders have complex sets of neural correlates, whose properties appear to go beyond simple hyper/hypoconnectivity, and are perhaps better described by differences in network shape rather than connection strength. The framework of network integration and segregation provides an easily implemented and easily understood set of common tools that researchers can use to better characterize the complex pattern of differences often observed in neuroimaging studies of individuals with these disorders. Adding this framework to the repertoire of neuroimaging research adds a powerful tool for understanding the complex nature of neurodevelopmental and other disorders.

REFERENCES

Abbott, A.E., Nair, A., Keown, C.L., Datko, M., Jahedi, A., Fishman, I., Müller, R.-A., 2016. Patterns of atypical functional connectivity and behavioral links in autism differ between default, salience, and executive networks. Cereb. Cortex 26 (10), 4034–4045.

Achard, S., Salvador, R., Whitcher, B., Suckling, J., Bullmore, E., 2006. A resilient, low-frequency, small-world human brain functional network with highly connected association cortical hubs. J. Neurosci. 26 (1), 63–72.

Albert, R., Jeong, H., Barabási, A.L., 2000. Error and attack tolerance of complex networks. Nature 406 (6794), 378–382.

Alexander, G., 1986. Parallel organization of functionally segregated circuits linking basal ganglia and cortex. Annu. Rev. Neurosci. 9 (1), 357–381.

Alexander, G.E., Crutcher, M.D., 1990. Functional architecture of basal ganglia circuits: neural substrates of parallel processing. Trends Neurosci. 13 (7), 266–271.

American Psychiatric Association, 2013. Diagnostic and Statistical Manual of Mental Disorders, fifth ed. American Psychiatric Publishing, Arlington, VA.

Baron-Cohen, S., Ring, H.A., Bullmore, E.T., Wheelwright, S., Ashwin, C., Williams, S.C.R., 2000. The amygdala theory of autism. Neurosci. Biobehav. Rev. 24 (3), 355–364.

Beare, R., Adamson, C., Bellgrove, M.A., Vilgis, V., Vance, A., Seal, M.L., Silk, T.J., 2016. Altered structural connectivity in ADHD: a network based analysis. Brain Imaging Behav. 11 (3), 846–858.

Betzel, R.F., Fukushima, M., He, Y., Zuo, X.N., Sporns, O., 2016. Dynamic fluctuations coincide with periods of high and low modularity in resting-state functional brain networks. NeuroImage 127, 287–297.

Brier, M.R., Thomas, J.B., Fagan, A.M., Hassenstab, J., David, M., 2014. Functional connectivity and graph theory in preclinical Alzheimer's disease. Neurobiol. Aging 35 (4), 1–25.

Bullmore, E.T., Sporns, O., 2009. Complex brain networks: graph theoretical analysis of structural and functional systems. Nat. Rev. Neurosci. 10 (3), 312.

Calhoun, V.D., Miller, R., Pearlson, G., Adali, T., 2014. The chronnectome: time-varying connectivity networks as the next frontier in fMRI data discovery. Neuron 84 (2), 262–274.

Callaway, D.S., Newman, M.E.J., Strogatz, S.H., Watts, D.J., 2000. Network robustness and fragility: percolation on random graphs. Phys. Rev. Lett. 85 (25), 5468–5471.

Cao, Q., Shu, N., An, L., Wang, P., Sun, L., Xia, M.-R., Wang, J.H., Gong, G.L., Zang, Y.F., Wang, Y.F., He, Y., 2013. Probabilistic diffusion Tractography and graph theory analysis reveal abnormal white matter structural connectivity networks in drug-naive boys with attention deficit/hyperactivity disorder. J. Neurosci. 33 (26), 10676–10687.

Cao, M., Shu, N., Cao, Q., Wang, Y., He, Y., 2014. Imaging functional and structural brain connectomics in attention-deficit/hyperactivity disorder. Mol. Neurobiol. 50 (3), 1111–1123.

Castellanos, F.X., Aoki, Y., 2016. Intrinsic functional connectivity in attention-deficit/hyperactivity disorder: a science in development. Biol. Psych. Cogn. Neurosci. Neuroimag. 1 (3), 253–261.

Center for Disease Control, 2018. Key Findings: National Prevalence of ADHD and Treatment: New statistics for children and adolescents, 2016.

Cerliani, L., Mennes, M., Thomas, R.M., Di Martino, A., Thioux, M., Keysers, C., 2015. Increased functional connectivity between subcortical and cortical resting-state networks in autism spectrum disorder. JAMA Psychiat. 72 (8), 767–777.

Cohen, J.R., 2017. The behavioral and cognitive relevance of time-varying, dynamic changes in functional connectivity. NeuroImage. (In Press, Available Online).

Cohen, J.R., D'Esposito, M., 2016. The segregation and integration of distinct brain networks and their relationship to cognition. J. Neurosci. 36 (48), 12083–12094.

Cortese, S., Kelly, C., Chabernaud, C., Proal, E., Di Martino, A., Milham, M.P., Castellanos, F.X., 2012. Toward systems neuroscience of ADHD: a meta-analysis of 55 fMRI studies. Am. J. Psychiatry 169 (10), 1038–1055.

Cubillo, A., Halari, R., Smith, A., Taylor, E., Rubia, K., 2012. A review of fronto-striatal and fronto-cortical brain abnormalities in children and adults with attention deficit hyperactivity disorder (ADHD) and new evidence for dysfunction in adults with ADHD during motivation and attention. Cortex 48 (2), 194–215.

Deco, G., Kringelbach, M.L., 2014. Great expectations: using whole-brain computational connectomics for understanding neuropsychiatric disorders. Neuron 84 (5), 892–905.

Deco, G., Tononi, G., Boly, M., Kringelbach, M.L., 2015. Rethinking segregation and integration: contributions of whole-brain modelling. Nat. Rev. Neurosci. 16 (7), 430–439.

Di Martino, A., Ross, K., Uddin, L.Q., Castellanos, F.X., Sklar, A.B., Milham, M.P., 2009. Functional brain correlates of social and nonsocial processes in autism spectrum disorders: an activation likelihood estimation meta-analysis. Biol. Psychiatry 65 (1), 63–74.

Di Martino, A., Kelly, C., Grzadzinski, R., Zuo, X.N., Mennes, M., Mairena, M.A., Lord, C., Castellanos, F.X., Milham, M.P., 2011. Aberrant striatal functional connectivity in children with autism. Biol. Psychiatry 69 (9), 847–856.

Di Martino, A., Fair, D.A., Kelly, C., Satterthwaite, T.D., Castellanos, F.X., Thomason, M.E., Craddock, R.C., Luna, B., Leventhal, B.L., Zuo, X., Milham, M.P., 2014. Unraveling the miswired connectome: a developmental perspective. Neuron 83 (6), 1335–1353.

Fair, D.A., Posner, J., Nagel, B.J., Bathula, D., Costa-Dias, T.G., Mills, K.L., Blythe, M.S., Giwa, A., Schmitt, C.F., Joel, T., 2010. Atypical default network connectivity in youth with ADHD. Biol. Psychiatry 68 (12), 1084–1091.

Fair, D.A., Nigg, J.T., Iyer, S., Bathula, D., Mills, K.L., Dosenbach, N.U.F., Schlaggar, B.L., Mennes, M., Gutman, D., Bangaru, S., Buitelaar, J.K., Dickstein, D.P., Di Martino, A., Kennedy, D.N., Kelly, C., Luna, B., Schweitzer, J.B., Velanova, K., Wang, Y.F., Mostofsky, S., Castellanos, F.X., Milham, M.P., 2013. Distinct neural signatures detected for ADHD subtypes after controlling for micro-movements in resting state functional connectivity MRI data. Front. Syst. Neurosci. 6, .

Fornito, A., Zalesky, A., Breakspear, M., 2015. The connectomics of brain disorders. Nat. Rev. Neurosci. 16 (3), 159–172.

Freeman, L.C., 1978. Centrality in social networks conceptual clarification. Soc. Netw. 1 (3), 215–239.

Green, S.A., Hernandez, L., Bookheimer, S.Y., Dapretto, M., 2016. Salience network connectivity in autism is related to brain and behavioral markers of sensory overresponsivity. J. Am. Acad. Child Adolesc. Psychiatry 55 (7), 618–626.e1.

Hagerman, P.J., Hagerman, R.J., 2004. The fragile-X premutation: a maturing perspective. Am. J. Hum. Genet. 74 (5), 805–816.

Hawellek, D.J., Hipp, J.F., Lewis, C.M., Corbetta, M., Engel, A.K., 2011. Increased functional connectivity indicates the severity of cognitive impairment in multiple sclerosis. Proc. Natl. Acad. Sci. 108 (47), 19066–19071.

Henry, T.R., Dichter, G.S., Gates, K., 2017. Age and gender effects on intrinsic connectivity in autism using functional integration and segregation. Biol. Psych. Cogn. Neurosci. Neuroimag. 3 (5), 414–422.

Holland, P.W., Leinhardt, S., 1971. Transitivity in structural models of small groups. Comp. Group Stud. 2, 107–124.

Hull, J.V., Jacokes, Z.J., Torgerson, C.M., Irimia, A., Van Horn, J.D., 2016. Resting-state functional connectivity in autism spectrum disorders: a review. Front. Psych. 7.

Itahashi, T., Yamada, T., Watanabe, H., Nakamura, M., Jimbo, D., Shioda, S., Toriizuka, K., Kato, N., Hashimoto, R., 2014. Altered network topologies and hub organization in adults with autism: a resting-state fMRI study. PLoS One 9 (4), e94115.

Kana, R.K., Uddin, L.Q., Kenet, T., Chugani, D., Mueller, R.-A., 2014. Brain connectivity in autism. Front. Hum. Neurosci. 8.

Keown, C.L., Datko, M.C., Chen, C.P., Maximo, J.O., Jahedi, A., Müller, R.-A., 2017. Network organization is globally atypical in autism: a graph theory study of intrinsic functional connectivity. Biol. Psych. Cogn. Neurosci. Neuroimag. 2 (1), 66–75.

Konrad, K., Eickhoff, S.B., 2010. Is the ADHD brain wired differently? A review on structural and functional connectivity in attention deficit hyperactivity disorder. Hum. Brain Mapp. 31 (6), 904–916.

Kucyi, A., Tambini, A., Sadaghiani, S., Keilholz, S., Cohen, J.R., 2018. Spontaneous cognitive processes and the behavioral validation of time-varying brain connectivity. Netw. Neurosci. (In Press, Available Online).

Latora, V., Marchiori, M., 2001. Efficient behavior of small-world networks. Phys. Rev. Lett. 87 (19), 198701.

Lewis, J.D., Theilmann, R.J., Townsend, J., Evans, A.C., 2013. Network efficiency in autism spectrum disorder and its relation to brain overgrowth. Front. Hum. Neurosci. 7.

Lin, P., Sun, J., Yu, G., Wu, Y., Yang, Y., Liang, M., Liu, X., 2014. Global and local brain network reorganization in attention-deficit/hyperactivity disorder. Brain Imaging Behav. 8 (4), 558–569.

Martínez-Pedraza, F.d.L., Carter, A.S., 2009. Autism spectrum disorders in young children. Child Adolesc. Psychiatr. Clin. N. Am. 18 (3), 645–663.

Menon, V., 2015. Salience network. In: Toga, A.W. (Ed.), Brain Mapping: An Encyclopedic Reference, vol. 2. Academic Press, pp. 597–611.

Menon, V., Uddin, L.Q., 2010. Saliency, switching, attention and control: a network model of insula function. Brain Struct. Funct. 1–13.

Mueller, R.A., Shih, P., Keehn, B., Deyoe, J.R., Leyden, K.M., Shukla, D.K., 2011. Underconnected, but how? A survey of functional connectivity MRI studies in autism spectrum disorders. Cereb. Cortex 21 (10), 2233–2243.

Newman, M.E.J., 2006. Modularity and community structure in networks. Proc. Natl. Acad. Sci. 103 (23), 8577–8582.

Odriozola, P., Uddin, L.Q., Lynch, C.J., Kochalka, J., Chen, T., Menon, V., 2016. Insula response and connectivity during social and non-social attention in children with autism. Soc. Cogn. Affect. Neurosci. 11 (3), 433–444.

Padmanabhan, A., Lynch, C.J., Schaer, M., Menon, V., 2017. The default mode network in autism. Biol. Psych. Cogn. Neurosci. Neuroimag. 2 (6), 476–486.

Plichta, M.M., Scheres, A., 2014. Ventral-striatal responsiveness during reward anticipation in ADHD and its relation to trait impulsivity in the healthy population: a meta-analytic review of the fMRI literature. Neurosci. Biobehav. Rev. 38, 125–134.

Posner, J., Park, C., Wang, Z., 2014. Connecting the dots: a review of resting connectivity MRI studies in attention-deficit/hyperactivity disorder. Neuropsychol. Rev. 24 (1), 3–15.

Ray, S., Miller, M., Karalunas, S., Robertson, C., Grayson, D.S., Cary, R.P., Hawkey, E., Painter, J.G., Kriz, D., Fombonne, E., Nigg, J.T., Fair, D.A., 2014. Structural and functional connectivity of the human brain in autism spectrum disorders and attention-deficit/hyperactivity disorder: a rich club-organization study. Hum. Brain Mapp. 35 (12), 6032–6048.

Roine, U., Roine, T., Salmi, J., Nieminen-von Wendt, T., Tani, P., Leppämäki, S., Rintahaka, P., Caeyenberghs, K., Leemans, A., Sams, M., 2015. Abnormal wiring of the connectome in adults with high-functioning autism spectrum disorder. Mol. Autism 6 (1), 65.

Rudie, J.D., Brown, J.a., Beck-Pancer, D., Hernandez, L.M., Dennis, E.L., Thompson, P.M., Bookheimer, S.Y., Dapretto, M., 2013. Altered functional and structural brain network organization in autism. NeuroImage Clin. 2, 79–94.

Seeley, W.W., Menon, V., Schatzberg, A.F., Keller, J., Glover, G.H., Kenna, H., Reiss, A.L., Greicius, M.D., 2007. Dissociable intrinsic connectivity networks for salience processing and executive control. J. Neurosci. 27 (9), 2349–2356.

Shine, J.M., Poldrack, R.A., 2017. Principles of dynamic network reconfiguration across diverse brain states. NeuroImage . (In Press, Available Online).

Sonuga-Barke, E.J.S., 2002. Psychological heterogeneity in AD/HD—a dual pathway model of behaviour and cognition. Behav. Brain Res. 130 (1), 29–36.

Sonuga-Barke, E.J.S., 2005. Causal models of attention-deficit/hyperactivity disorder: From common simple deficits to multiple developmental pathways. Biol. Psychiatry 57 (11), 1231–1238.

Sonuga-Barke, E.J.S., Castellanos, F.X., 2007. Spontaneous attentional fluctuations in impaired states and pathological conditions: a neurobiological hypothesis. Neurosci. Biobehav. Rev. 31 (7), 977–986.

Sporns, O., 2010. Networks of the Brain. MIT Press, Cambridge, MA.

Sporns, O., 2013. Network attributes for segregation and integration in the human brain. Curr. Opin. Neurobiol. 23 (2), 162–171.

Sporns, O., Betzel, R.F., 2016. Modular brain networks. Annu. Rev. Psychol. 67 (1), 613–640.

Uddin, L.Q., Menon, V., 2009. The anterior insula in autism: under-connected and under-examined. Neurosci. Biobehav. Rev. 33 (8), 1198–1203.

Uddin, L.Q., Kelly, A.M.C., Biswal, B.B., Margulies, D.S., Shehzad, Z., Shaw, D., Ghaffari, M., Rotrosen, J., Adler, L.A., Castellanos, F.X., Milham, M.P., 2008. Network homogeneity reveals decreased integrity of default-mode network in ADHD. J. Neurosci. Methods 169 (1), 249–254.

Uddin, L.Q., Supekar, K., Lynch, C.J., Khouzam, A., Phillips, J., Feinstein, C., Riyali, S., Menon, V., 2013. Salience network-based classification and prediction of symptom severity in children with autism. JAMA Psychiat. 70 (8), 869–879.

Uddin, L.Q., Dajani, D.R., Voorhies, W., Bednarz, H., Kana, R.K., 2017. Progress and roadblocks in the search for brain-based biomarkers of autism and attention-deficit/hyperactivity disorder. Transl. Psychiatry 7 (8), e1218.

van den Heuvel, M.P., Sporns, O., 2011. Rich-club organization of the human connectome. J. Neurosci. 31 (44), 15775–15786.

Wang, L., Zhu, C., He, Y., Zang, Y., Cao, Q., Zhang, H., Zhong, Q., Wang, Y., 2009. Altered small-world brain functional networks in children with attention-deficit/hyperactivity disorder. Hum. Brain Mapp. 30 (2), 638–649.

Weissman, D.H., Roberts, K.C., Visscher, K.M., Woldorff, M.G., 2006. The neural bases of momentary lapses in attention. Nat. Neurosci. 9 (7).

Addiction: Informing drug abuse interventions with brain networks

Vaughn R. Steele, Xiaoyu Ding, Thomas J. Ross

*Neuroimaging Research Branch, National Institute on Drug Abuse, Intramural Research Program,
National Institutes of Health, Baltimore, MD, United States*

CHAPTER OUTLINE

On the face of it, the chronic relapsing brain disease of addiction seems to be a perfect model system for studying brain networks. A large corpus of preclinical research has elucidated circuits known to be involved in most, if not all, abused substances. The disease cycle of addiction (see Fig. 1; Koob and Volkow, 2016; Volkow et al., 2016) is largely derived from dopaminergic dysfunction, (Volkow and Morales, 2015) specifically in dopamine (DA) released from the mesencephalic ventral tegmental area (VTA) into the nucleus accumbens (NAcc), prefrontal cortex (PFC), and amygdala. Abused substances differentially affect similar neural circuitry as primary, biological rewards such as food, water, and sex (Di Chiara, 2002; Wise, 2002). Drug use increases DA release in the mesocorticolimbic (MCL)-DA system (Kelley, 2004; Nestler, 2005), which is thought to be an important element in learning, goal-directed behavior, and reward processing (Everitt and Robbins, 2005; Kalivas and O'Brien, 2008) leading to drug dependence.

Although preclinical models give us a rich understanding of the circuits related to addiction, little has been translated to clinical practice in any meaningful way. We do know, broadly, individuals with substance use disorders (SUDs), relative to those without, exhibit dysregulation in attention, memory, reward processing, and executive control (Koob and Volkow, 2016; Spronk et al., 2013; Steele et al., 2017). Also, DA circuit dysregulations are linked to initiation and maintenance of addictive

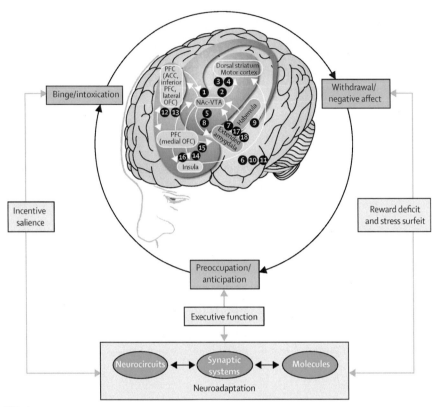

FIG. 1

The neural circuits that interact and are dysregulated in drug addiction. The stages of the addiction cycle include (1) binge and intoxication *(blue)*, (2) withdrawal and negative effect *(red)*, (3) and preoccupation and anticipation *(green)*. Arrows depict major circuit connections between domains, and numbers refer to neurochemical- and neurocircuit-specific pathways known to support brain changes that contribute to addiction. *ACC*, anterior cingulate cortex; *NAc-VTA*, nucleus accumbens-ventral tegmental area; *OFC*, orbitofrontal cortex; *PFC*, Prefrontal cortex.

From Koob, G.F., Volkow, N.D., 2016. Neurobiology of addiction: a neurocircuitry analysis. Lancet Psych. 3, 760–773. https://doi.org/10.1016/s2215-0366(16)00104-8.

behaviors (Goldstein and Volkow, 2002). A wide network of regions with connections to the MCL-DA are known to be dysregulated in SUDs such as the dorsolateral prefrontal cortex (dlPFC), anterior cingulate cortex (ACC), inferior frontal gyrus (IFG), orbitofrontal cortex (OFC), striatum, hippocampus, basolateral amygdala, and insula (for review, see Koob and Volkow, 2016; Spronk et al., 2013; Steele et al., 2017). Additionally, interconnections, or network connectivity, among these areas have been found to be dysregulated in SUD (Filbey et al., 2014; Gu et al., 2010; Kelly et al., 2011; Worhunsky et al., 2013).

Addiction, as a disease, is often characterized as an imbalance between reward (stronger) and executive control (weaker) networks (Bechara, 2005; Bickel et al., 2007). Considering addiction is often understood to be a disease of DA dysregulation (Volkow and Morales, 2015), regions within and connected to the MCL-DA are often interrogated. For example, reward processing is linked to the basal ganglia (Haber and Knutson, 2010), and functional differences are often characterized between individuals with and without a SUD. Although DA is often implicated in dysregulations apparent in addiction, thorough investigation of these mechanisms has yielded little in the way of successful clinical interventions (Nutt et al., 2015). This suggests moving beyond a singular focus of DA to a broad focus may be necessary to delineate appropriate interventions. A network-based understanding of the disease provides such a broad focus to better understand, diagnose, and treat the disease.

As an alternative to exploring reward-related networks, three networks and their interconnections within the MCL-DA system have emerged as primary networks of interest: the salience network (SN; Seeley et al., 2007), executive control networks (ECN; Seeley et al., 2007), and the default mode network (DMN; Raichle, 2015). The SN overlaps with reward circuitry but consists primarily of ACC and bilateral anterior insula (aI). As its name suggests, this network is associated with monitoring the salience of both exogenous and endogenous states (Seeley et al., 2007). The ECN consists primarily of dorsolateral frontal and parietal cortices (Seeley et al., 2007) and is often referred to as the task-positive network. Activation in the SN is thought to reflect bottom-up (i.e., automatic, stimulus driven) processes, and the ECN is thought to reflect top-down (i.e., controlled, internally driven) processes. In contrast, the DMN consists (primarily) of nodes in the medial PFC, the posterior cingulate cortex (PCC), and bilateral angular gyrus. This network is commonly found to deactivate during goal-directed behavior (Fox et al., 2005) and is frequently referred to as the task-negative network (see Fig. 2 for a representation of the three networks).

Alterations in the dynamic integration of these three networks are implicated in addiction. The SN appears to have a critical role interfacing between the DMN and the ECN (Sridharan et al., 2008; Menon and Uddin, 2010). The insula in particular has been demonstrated to have a critical role in addiction (Naqvi and Bechara, 2009), as cigarette smokers suffering insula-centered lesions due to stroke spontaneously lost their desire to smoke (Naqvi et al., 2007). Increased activation in the ECN is often accompanied with increased deactivation of the DMN (Fox et al., 2005; McKiernan et al., 2003). The ECN is implicated in many cognitive constructs, including maintenance of attention. Individuals with SUDs have exhibited deficiencies in both the ECN and SN as measured by processing drug-related cues (i.e., craving: SN; Garavan et al., 2000; Kühn and Gallinat, 2011) and executive control tasks (i.e., ECN; Hester, 2004; Kaufman et al., 2003). Among the many functions attributed to the DMN, internal ruminations/thinking about one's self and one's internal state is probably the one most implicated in drug abuse. A proposed biomarker for substance abuse is dysregulation between the SN and the DMN (Sutherland et al., 2012).

In this chapter, we review techniques used to understand human brains that are structurally and functionally organized into complex networks for segregating and

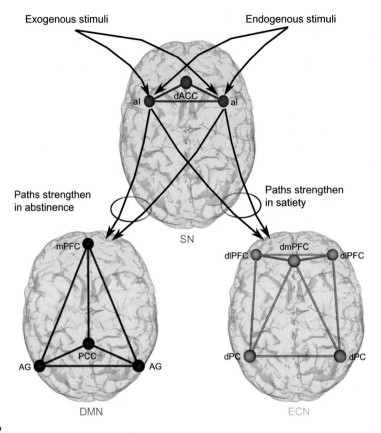

FIG. 2

The tripartite model (Sutherland et al. 2012) of activity within and between salience network (SN), executive control network (ECN), and the default mode network (DMN). The insula, in the SN, is posited to be the switch between endogenously and exogenously salient events. In the abstinent state, stronger coupling exists between the SN and the DMN, leading to an overemphasis on internal states (e.g., craving and its physiological manifestations). In the sated state, there is greater network balance and the SN can engage the ECN as necessary to focus attention on external stimuli.

integrating information to define brain networks using functional connectivity (FC) in SUD populations. It is essential to understand how drugs affect the topological organization of brain connectivity networks to diagnose and treat individuals suffering from SUDs. In recent years, network connectivity has been studied using noninvasive brain imaging techniques, in particular functional magnetic resonance imaging (fMRI), to extract connectivity measures during resting state. There are several methods we will discuss later to analyze resting-state functional connectivity (rsFC) and task-based functional connectivity (tbFC). The studies discussed later provided sufficient information that drugs change regional and large-scale brain network functions and that those changes can predict drug-related outcomes (Steele et al., 2018; Adinoff et al., 2015; McHugh et al., 2016). The emerging view from the clinical literature is that drug abuse involves several brain processes that cannot be ascribed to single brain regions or neurotransmitter systems but rather are a consequence of brain networks and their interactions. We describe addiction as a brain-networks disease, with an emphasis on reward processing, SN, ECN, and DMN. Finally, we speculate on how these results can inform future clinical treatment.

SEED-BASED FUNCTIONAL CONNECTIVITY

The initial and perhaps most widely used technique to explore FC is a seed-based analysis. In this analysis, a seed region is defined *a priori*, and a correlation is calculated between the seed and other voxels to generate a FC map. This FC correlation strength is interpreted as a measure of the network strength between the seed to other regions. For drug addiction, many of these seeds originate in the reward system or within one of the three aforementioned networks (i.e., DMN, SN, or ECN).

Seed-based analyses often highlight an imbalance within users who exhibit stronger reward and weaker ECNs (Bechara, 2005; Bickel et al., 2007). For example, compulsive drug use is linked to dysregulation of brain circuits related to "Go" and "Stop" processes. Dysregulation of the "Go" circuit, FC between the right ventral striatum and superior-anterior PFC/OFC, is thought to promote compulsive drug use (Hu et al., 2015). Dysregulation of the "Stop" circuit, right ventral striatum and inferior-dorsal ACC (dACC), may limit compulsive drug use (Hu et al., 2015).

The reward system is a complex network of regions (Steele et al., 2017) that together learn associations between goals and rewards with phasic and tonic changes within the DA-MCL system (Everitt and Robbins, 2005). Reward processing is thought to be most influential to the disease process during initiation and maintenance (i.e., regular use) of drug use, and to a lesser extent in the abuse and addiction part of the cycle (Koob and Volkow, 2016; Volkow et al., 2016). The basal ganglia are heavily implicated in reward processing (Haber and Knutson, 2010) and are often used as a seed in FC analysis. For example, heroin users (HU), relative to healthy controls (HC), exhibit greater FC between a seed in the NAcc and ventral/rostral ACC (v/rACC) and OFC (Ma et al., 2010). Plasticity in these networks is also identified by the negative relation between decreased FC strength from the VTA to thalamus/lentiform nucleus/NAcc and years of cocaine use (Gu et al., 2010). Another reward seed, bilateral putamen, identified cocaine users (CU) to have less tbFC between the putamen and both the posterior insula (pI) and right postcentral gyrus (McHugh et al., 2013). In CU, impulsivity is correlated with increased FC between a striatal seed and dlPFC (Hu et al., 2015) and lower tbFC between putamen and pI (McHugh et al., 2013). Risk of relapse to cocaine is increased with greater impulsivity and reduced tbFC between the putamen and pI/postcentral gyrus (McHugh et al., 2013).

Two nodes of the SN are common seeds in FC analyses. First, ACC seeds identify connectivity related to nicotine users (NU; Hong et al., 2009, 2010). A dACC seed identified FC with striatal regions related to nicotine addiction, which can be enhanced with nicotine administration (Hong et al., 2009). dACC rsFC with right ventral striatum is lower in smokers compared with nonsmokers (Hong et al., 2010). FC between dACC and bilateral ventral striatum predicts nicotine addiction severity with a genetic component (α5 Asp398Asn polymorphism; "risk allele") accounting for up to 12.8% of the variance of smoking (Hong et al., 2010), suggesting a genetic biomarker for a predisposition for nicotine addiction. Second, insular seeds identify connectivity related to NU (Fedota et al., 2016) and CU (Cisler et al., 2013). The insula has been an important node of the SN that informs addiction because focal lesion of the area resulted in spontaneous smoking cessation (Naqvi et al., 2007).

A node of the SN, the insula, has also been identified with either task-defined (Fedota et al., 2016) or independent component analysis (ICA)-defined (Cisler et al., 2013) methods. Smokers, relative to nonsmokers, exhibited weaker FC from the task-defined right insula to superior frontal gyrus (Fedota et al., 2016). An insular seed defined using resting-state ICA showed CU, relative to controls, exhibited greater connectivity in right insula with dorsomedial PFC, IFG, and bilateral dlPFC (Cisler et al., 2013).

Some researchers use the PCC as seed within the DMN to identify FC in addiction. First, greater basal FC was identified between the PCC and left posterior hippocampus in CU who relapsed after treatment, compared with those did not relapse (Adinoff et al., 2015). Second, individuals who were cannabis-dependent, nicotine-dependent, or both showed lower FC between the PCC and medial frontal cortex, inferior parietal cortex, and temporal cortex (nodes of the DMN), compared with HC (Wetherill et al., 2015). The PCC was also shown to have lower FC with the cerebellum, medial PFC (mPFC), parahippocampus, and aI in cannabis- and nicotine-dependent individuals (Wetherill et al., 2015). Finally, duration of cannabis use was positively correlated with FC between PCC and right aI (Wetherill et al., 2015) suggesting neuroplastic changes related to long-term use.

Beyond the aforementioned networks, additional seeds-based regions have been used in addiction research. In HU, FC is lower between PFC and OFC, and between PFC and ACC (Ma et al., 2010). FC with bilateral amygdala as a seed predicted relapse in CU (McHugh et al., 2014). Lower FC between the left corticomedial amygdala and ventromedial PFC/rACC was identified in relapsed CU relative to nonrelapsed CU after treatment. CU who did not relapse exhibited lower FC between bilateral basolateral amygdala and lingual gyrus/cuneus relative to relapsed CU and HC. Therefore, amygdala FC may be a biomarker for relapse risk in CU.

INDEPENDENT COMPONENT ANALYSIS

There are several data-reduction techniques used to define networks such as principle component analysis or nonnegative matrix factorization, but ICA (for review, see Calhoun et al., 2009) has been the most successful approach with neuroimaging. Spatial ICA, as most commonly applied to fMRI, seeks to decompose a matrix (time by voxels) into the product of two matrices (time by components and components by voxels), subject to the constraint that the columns in the components by voxels matrix are maximally statistically independent. Having the number of components N be less than the number of time points results in data reduction and N-independent spatial maps with their corresponding (nonindependent) time courses. Although there are principled ways to determine an optimum N (Abou-Elseoud et al., 2010; Li et al., 2007), practically most researchers arbitrarily choose a number in the 20–40 range.

Mckeown et al. (1998) initially applied ICA to fMRI task data. However, Smith et al. (2009) paved the way for using rsFC and ICA to define brain networks. They performed ICA on a large database of task-activation locations and showed that the resulting networks were similar to those obtained by ICA on resting-state data. Thus, one can probe these networks in the absence of obtaining task data in difficult-to-scan populations such as, in our case, drug-using groups who may perform tasks poorly or be generally amotivational. Critically, these networks have been shown to be consistent (Damoiseaux et al., 2006), stable over time (Chen et al., 2008), and are thus suitable for use in drug trials using crossover designs and longitudinal treatment studies. The SN, ECN, and DMN implicated in drug abuse are also consistent, although with higher N ICA decomposition, these three networks will split into smaller subnetworks (i.e., typically the ECN will split into left- and right-lateralized networks, and DMN will split into an anterior and posterior network; Abou-Elseoud et al., 2010).

Using ICA on rsFC to measure FC and diffusion tensor imaging (DTI) to measure white-matter structural differences, Ma et al. (2015) found numerous DMN deficits in HU compared with HC. HU had lower DMN connectivity in the OFC, the bilateral inferior parietal lobule, the bilateral superior frontal gyrus, the bilateral paracentral lobule, the left parahippocampal gyrus (l-PHG), and the right middle temporal gyrus. Using fiber tracking between DMN nodes, they found that HU had higher mean diffusivity and lower fractional anisotropy (FA) between the PCC and PHG, and lower FA and track count between PCC and mPFC. Critically, white-matter differences between PCC and r-PHG, and PCC and mPFC, were negatively related to duration of heroin use. With ICA-defined DMN in HU undergoing methadone maintenance treatment, Li et al. (2015) showed lower connectivity in the left inferior temporal gyrus (l-ITG) and right superior occipital gyrus, with greater connectivity in the left precuneus and right middle cingulum in individuals who subsequently relapsed. Of note, the l-ITG region had a negative correlation with the number of relapse events (i.e., the greater number of relapses, the lower connectivity between DMN and l-ITG).

Acute nicotine withdrawal has also been related to DMN dysfunction. Cole et al. (2010) used ICA on rsFC to assess interactions between ECN and DMN in smokers who were abstinent, using nicotine gum, or placebo in a crossover design. Between-network coupling was positively correlated with changes in a withdrawal measurement of difficulty concentrating. They also performed a voxel-wise comparison within networks and withdrawal scores. Within the ECN, withdrawal symptom reductions were correlated to changes including to the left OFC, left dlPFC, and left thalamus, whereas the opposite was found in areas including the right pI and right MFG. For the DMN, symptom reductions were correlated to changes in regions including the left midcingulate and left OFC, with left dorsal hippocampus, bilateral precuneus, and PHG showing the opposite relationship.

In a study combining both task and rsFC, Fedota et al. (2016) used a parametric flanker task (Forster et al., 2010) in smokers and controls, and found differences in an ICA-defined right aI for medium difficulty trials. Using this node of the SN as a seed region, they examined rsFC differences from there to the SN, ECN, and DMN. They found that smokers had lower connectivity to the right superior frontal gyrus in the ECN.

In a combined functional and effective connectivity study, Ding and Lee (2013) used ICA to identify SN, ECN, and anterior and posterior DMN networks, testing each network's connections to the rest of the brain in abstinent and sated smokers. With a few notable exceptions, they found abstinent smokers had greater connectivity from those networks, mostly to other within-network nodes. Of particular note was greater connectivity in abstinence from the SN to right aI, and from the DMN to dACC, a SN node. Eight regions showing condition differences within hypothesized regions were used as regions of interest (ROIs) in an effective connectivity analysis using Granger causality. A complex pattern emerged but one that seems to show greater insular connections in abstinence between ACC, dlPFC, and precuneusand greater connectivity in the sated state between ACC and dlPFC, dACC, and PHG.

Cocaine relapse was investigated in the three-network model (McHugh et al., 2016). This group compared rsFC in ICA-defined SN, ECN (left and right), and DMN in CU posttreatment comparing those who did ($N = 24$) or did not ($N = 21$) relapse at 1 month. Abstinent individuals had greater connectivity between the left ECN and right middle and superior frontal gyrii, and right dACC; and between right ECN and left middle temporal gyrus (MTG). Critically, these "neuroprotective" effects remained for 6 months in the eight individuals who remained abstinent at that point.

Results such as those previously outlined led members of our group to propose a tripartite model of (specifically nicotine) addiction (see Fig. 2; Sutherland et al., 2012). In this model, the insula in the SN is posited to be the switch between endogenously and exogenously salient events. In the abstinent state, stronger coupling exists between the SN and the DMN, leading to an overemphasis on internal states (e.g., craving and its physiological manifestations). In the sated state, there is greater network balance, and the SN can engage the ECN as necessary to focus attention on external stimuli. In a direct test of this hypothesis, a measure of interaction between these networks, the resource allocation index (RAI, the difference in z-transformed coupling between SN and ECN compared with SN and DMN) was proposed (Lerman et al., 2014). In this work, 54 smokers had rsFC scans in the abstinent and sated conditions. ICA was used to define the four networks (SN, left- and right-ECN, and DMN), and the time-courses of these networks were correlated to compute the RAI. We found that the RAI was lower in abstinence, consistent with the proposed model (Sutherland et al., 2012). The RAI measure was also negatively correlated with a craving measure (urge to smoke).

We note that, although much of the field has focused on three networks and we thus focused our review therein, there have been many notable exceptions. For example, Schmidt et al. (2015) directly probed the DA system by using an ICA-defined basal ganglia circuit in HU under an acute drug challenge (heroin/placebo crossover design). They found increased rsFC in left putamen after heroin injection with no difference between HU and HC.

GRAPH THEORY

Graph theory provides unique insights into the topological organization of the human brain. Brain networks are represented as a graph with N nodes and K edges. The nodes usually represent anatomically defined brain regions, and the edges indicate the functional interactions (e.g., correlation) between these regions. A graph can be classified as directed or undirected and unweighted (binary) or weighted (continuous) depending upon if the edges have direction or strength, respectively. Most studies of drug addiction adopt undirected, unweighted graphs in which correlation matrices are thresholded and converted into adjacency matrices representing which nodes are connected. Graph theory provides a range of measures to describe network topology allowing measurement of subtle drug effects. Furthermore, these measures quantify drug-induced changes in network organization and thus provide a biological explanation of behavior related to individuals with SUDs (Gießing and Thiel, 2012). Graph theory is usually applied to rsFC to investigate brain topologies and connectomes. However, due to the nature of graph analysis and its ability to analyze a complex network with many nodes, highlighting results in relation to the networks discussed previously is not practical. Therefore, results are discussed later insomuch as the complex networks identified overlap with reward processing, SN, ECN, and DMN.

Commonly used topological measures fall into three categories: small world and efficiency, modularity, and nodal properties. In small world and efficiency measures, shortest path length and clustering coefficient are the two most elementary properties of a network. The shortest path length is calculated by averaging the minimum number of connections that link each possible pair of nodes. Global efficiency, which is the average inverse shortest path length in the network, measures the capacity of a network for parallel information transfer and integration. The clustering coefficient is a ratio between 0 and 1 that defines the proportion of possible connections that exist between the nearest neighbors of a node (Watts and Strogatz, 1998). Local efficiency, closely related to the clustering coefficient, is the global efficiency computed on the neighborhood of the node. It reflects the capacity of the network for information transfer between the nearest neighbors of a particular node. A graph is considered small-world if its average clustering coefficient is significantly higher than a random graph constructed with the same number of nodes, and if the graph has a small average shortest path length. Modularity is another important measure for describing the organization of a network. A module refers to a group of nodes highly connected among themselves but less connected to the others within the networks (Newman, 2006). Modularity quantifies the efficacy of partitioning a network by evaluating the difference between the actual number of intramodule connections and the expected number for a random network. It reflects the degree of how a network is organized into a community structure. After module assignment, patterns of intramodule and intermodule connectivity at the module level and nodal level are usually examined (Guimerà and Amaral, 2005). The most commonly evaluated nodal properties of a network are nodal degree, nodal efficiency, and nodal betweenness centrality. The degree of a node, representing the most local and directly quantifiable measure of centrality, is defined as the number of edges that link to the node. The nodal efficiency is numerically the inverse of the averaged shortest path length between the node and each of the other nodes in the network, which reflects the ability of information transfer from the node to other nodes (Achard and Bullmore, 2007). The nodal betweenness centrality is the number of shortest paths between any two nodes that pass through the node.

HU showed a lower small-worldness compared with HC (Liu et al., 2009; Jiang et al., 2013). Also, they exhibit lower nodal centrality in regions of the ECN, including the bilateral middle cingulate gyrus, dlPFC, and right precuneus. Moreover, compared with controls, HU had greater nodal centrality in the left hippocampus, which was also positively correlated with their duration of heroin addiction (Jiang et al., 2013). HU also have abnormal topological properties, such as higher nodal degrees, in the PFC, ACC, supplementary motor area, ventral striatum, insula, amygdala, and hippocampus, linking topological properties with known behavioral manifestation of the disease including lower self-control, impaired inhibitory function, and deficits in stress regulation (Liu et al., 2009). This group related rsFC to regions related to activation elicited during heroin cue-reactivity task (Liu et al., 2011). The nodal degree of the MFG in HU rsFC was significantly higher and positively related to the neural activity in the same region in the cue-reactivity task. Furthermore, nodal degree in ACC in HU was negatively correlated with task-induced craving activation. Yuan et al. (2010) found abnormal topological properties in key addiction-related brain circuits (e.g., control, reward, motivation, and memory), that also revealed the duration of heroin use is positively correlated with the nodal degree in the r-PHG, left putamen, and bilateral cerebellum, as well as negatively correlated with the shortest path length in the same regions.

To investigate the relationship between cognitive performance and brain topological organization, Gießing et al. (2013) measured performance and brain activity in a Go/NoGo task and rsFC from abstinent smokers in a placebo-controlled, crossover design. They tested the main effects of nicotine and time-on-task on behavioral performance, and brain network topological properties such as efficiency, clustering, and connection distance. They found that nicotine enhanced attentional task performance and increased efficiency and connection distance, whereas it decreased clustering parameters of the brain networks. Lin et al. (2015) examined alterations in topological organization of rsFC in satiated heavy smokers. Compared with nonsmokers, heavy smokers showed greater nodal local efficiency predominantly in visual regions but lower nodal global efficiency in DMN areas, which were related to the duration of cigarette use or the severity of nicotine dependence. However, in another study, smoking status did not significantly change the efficiency of functional networks (Breckel et al., 2013). Both smokers and nonsmokers had similar regional and nodal network properties.

There are few studies applying graph theory to other drugs of abuse. Interactions between the SN and DMN were disrupted in CU (Liang et al., 2015). Specifically, CU have lower intermodule connectivity between DMN and SN at the module level, as well as lower connections between the rACC and SN, the PCC and ECN, and the bilateral insula and DMN at the nodal level. Individuals with alcohol dependence show significantly lower efficiency indices between the PCC and multiple cerebellar regions but higher efficiency in several regions correlated with length of sobriety (Chanraud et al., 2011). In contrast, during a working memory task, individuals with alcohol dependence, relative to HC, had more robust connectivity between the l-PCC and left cerebellar regions, suggesting a compensatory network to achieve normal performance. Also, individuals with alcohol dependence showed lower average cluster coefficients related to more severe alcohol use, whereas lower global efficiency, degree, and clustering coefficient were associated with longer alcohol dependence duration (Sjoerds et al., 2017). Furthermore, within four bilateral nodes in striatum, alcohol use severity was negatively correlated with the clustering coefficient in the left caudate, whereas alcohol dependence duration was negatively associated with the clustering coefficient in the caudate and putamen, and negatively related to the nodal degree in the bilateral caudate. Chronic polydrug users with a primary diagnosis of cocaine dependence showed, relative to controls, greater connection strength but lower communication efficiency and small-worldness, which indicates a loss of normal interregional communications and network topology features in the addicted brain (Wang et al., 2015).

Finally, we are aware of only one paper that used weighted networks to explore the effects of drugs on brain-network properties. They found that interconnections among communities are enhanced while under the influence of psilocybin (see Fig. 3; Petri et al., 2014). The dramatic change in the topological structure of the brain's functional patterns may reflect an increased integration among networks compared with placebo suggesting at least short-term malleability with networks related to addiction.

Generally, due to few studies and inconsistent results, strong conclusions related to drug addiction based on graph theory analysis alone prove difficult. Inconsistencies may be due to idiosyncrasies in the groups such as addiction severity or drug type. Nonetheless, graph theory applied to resting-state fMRI data then mapped back to task data may be most influential in the future. With this integration, both the identification of the complex network related to drug addiction and the underlying cognitive functions related to the behavioral manifestation of the disease are possible. Therefore, graph theory analysis could provide novel and theoretically important insights into the neurobiological mechanisms of drug addiction.

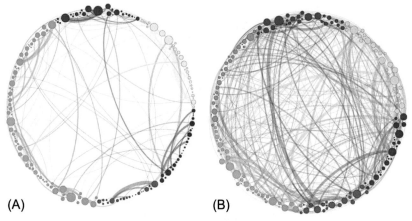

FIG. 3

Graph theory using weighted networks and a topological analysis applied to addiction. *Colors* denote communities, edge widths are proportional to the weight and node size is proportional to strength. Individuals under the influence of psilocybin (B) compared with placebo (A) showed heavier links between communities, which could reflect an increased integration among networks while under the influence of the drug.

From Petri, G., Expert, P., Turkheimer, F., Carhart-Harris, R., Nutt, D., Hellyer, P.J., Vaccarino, F., 2014. Homological scaffolds of brain functional networks. J. R. Soc. Interf. 11, 20140873. https://doi.org/10.1098/rsif.2014.0873.

DISCUSSION AND FUTURE DIRECTIONS

Large-scale brain networks, with an emphasis on reward processing, SN, ECN, and DMN, are useful for identifying differences between individuals with and without SUDs. Although much could yet be understood about these networks in addiction, current techniques appropriately applied can identify risk factors and positively affect development of new interventions.

Our group has made efforts to combine all brain networks in predictive models of drug abuse. The hope is that these models could one day lead to quantitative biomarkers of addiction. For example, we used 16 ICA components from an analysis of the brainmap database (Smith et al., 2009), leading to 56 network nodes. We then constructed three feature types: within node, between network, and correlation of nodes within a net, for a total of 56, 120, and 119 features respectively. Using support vector machine (SVM) learning with cross-validation, we were able to predict smoking status with 76%–78% accuracy, with 2 ECN circuits among the 13 that were specifically predictive of smoking status (Pariyadath et al., 2014). We are continuing this work, to optimize processing resting-state data to improve these models (Ding et al., 2017). In addition to classifying status, we have created machine-learning models to predict treatment outcomes. In this case, ICA maps from task data were used to construct a between-network correlation matrix. Using this matrix as features in an SVM framework, we were able to predict who would complete treatment with >80% accuracy (Steele et al., 2018), much better than models based upon clinical assessments alone (about 67% accurate). Alternative modalities such as event-related potentials (ERPs) are successfully employed to predict substance abuse outcomes (Fink et al., 2016; Marhe et al., 2013; Steele et al., 2014) with measures that could be generated from caudal and rostral regions of the ACC (Edwards et al., 2012) or the PCC (Agam et al., 2011), suggesting either the SN or DMN is predictive of the outcomes measured. Because localization of neural generators of ERPs proves difficult, collecting ERPs simultaneous with fMRI could unlock the temporal and spatial dynamics related to SUDs, thus identifying biomarkers for risk factors and targets for treatment.

Pharmacological interventions are another possibility in treating addiction. Such interventions are known to modulate network connectivity in HC by targeting DA reward system (Cole et al., 2013) or administering nicotine (Tanabe et al., 2011). Networks are modulated in addiction models with simply abstaining (Fedota et al., 2017; Sutherland et al., 2012, 2013b) or with administration of methylphenidate (Konova et al., 2013), varenicline (Sutherland et al., 2013a), or methadone (Li et al., 2015; Ma et al., 2011; Wang et al., 2016). Clearly the dysregulated networks of addiction are malleable and prime targets for interventions. However, pharmacological interventions affect neurotransmitters globally causing nonspecific, off-target modifications that lead to side effects. Considering current treatments for addiction have an abysmal success rate (the most successful treatment is for nicotine where abstinence rates posttreatment are below 12%; Schlam and Baker, 2013), an alternative intervention that allows focal modifications of a specific network, while not modulating other networks, is needed.

Noninvasive brain stimulation (NIBS) has shown promise with focal modulation and specifically targeting networks known to be dysregulated in addiction (Diana et al., 2017). Repetitive transcranial magnetic stimulation (rTMS), a type of NIBS, over the left dlPFC is FDA-approved to treat depression (George et al., 1995; Pascual-leone et al., 1996), with substantial evidence for reducing depression symptoms (Berlim et al., 2014). Left dlPFC stimulation induces broad activity changes (Fox et al., 2012a) leading to a potential intervention tool in disorders of the DA system, such as addiction (Feil and Zangen, 2010; Jansen et al., 2013). Network connectivity between the dlPFC and ACC is normalized with this treatment (Fox et al., 2012b), and increases in DA release in the caudate nucleus (Keck et al., 2002; Strafella et al., 2001) suggests targeted network malleability with NIBS stimulation. In fact, NIBS has successfully reduced craving for cocaine (see Fig. 4; Camprodon et al., 2007; Hanlon et al., 2015; Politi et al., 2008; Terraneo et al., 2016), nicotine (see Fig. 5; Li et al., 2013), alcohol (Mishra et al., 2010), and opiates (Sahlem et al., 2017). The network dysregulations reviewed here related to addiction and neural activity in the ACC, insula, or striatum could be modulated by left dlPFC NIBS to facilitate long-term positive outcomes.

Overall, network connectivity related to reward processing, SN, ECN, and DMN are prime targets for interventions designed to modulate dysregulated processing in SUDs. The field is correctly drifting toward a network understanding of addiction

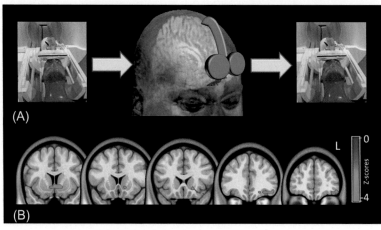

FIG. 4

Cortical magnetic stimulation modulates subcortical regions dysregulated in addiction. Cocaine participants received medial prefrontal continuous theta-burst stimulation (cTBS) and an fMRI. After cTBS, reductions were measured in cocaine craving and activation in the striatum. (A) Design and (B) TMS-evoked BOLD signal after cTBS.

From Hanlon, C.A., Dowdle, L.T., Austelle, C.W., DeVries, W., Mithoefer, O., Badran, B.W., George, M.S., 2015. What goes up, can come down: novel brain stimulation paradigms may attenuate craving and craving-related neural circuitry in substance dependent individuals. Brain Res. 1628, 199–209. https://doi.org/10.1016/j.brainres.2015.02.053.

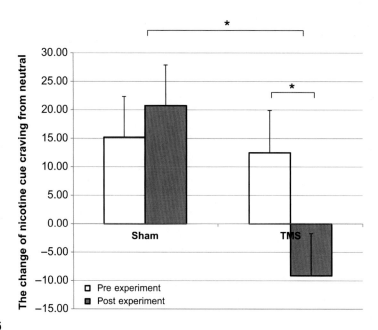

FIG. 5

An addiction intervention. In a sham-controlled procedure, transcranial magnetic stimulation (TMS) was administered over the left dorsolateral prefrontal cortex to nicotine users. Nicotine craving was reduced post-TMS treatment in the active TMS group but not the sham TMS group.

From Li, X., Hartwell, K.J., Owens, M., Lematty, T., Borckardt, J.J., Hanlon, C.A., Brady, K.T., George, M.S., 2013. Repetitive transcranial magnetic stimulation of the dorsolateral prefrontal cortex reduces nicotine cue craving. Biol. Psychiatry 73, 714–720. https://doi.org/10.1016/j.biopsych.2013.01.003.

rather than focusing on isolated regions. In the future, larger sample sizes are needed to better tease apart network connectivity differences related to drug abuse and controls, as well as identifying risk factors for poor outcomes. As the field moves toward individualized medicine, rich datasets that include several imaging modalities may be necessary to identify the most appropriate treatment for each individual (c.f., Drysdale et al., 2017). Such efforts must strike a balance by not including too many variables and run the risk of overfitting and thus developing a model that poorly generalizes to out-of-sample data. Multimodal datasets could yield a new understanding of risk factors and potentially new targets for intervention. NIBS is a recent SUD treatment possibility with focal manipulation of targeted dysregulated networks associated with SUDs. Nonetheless, the long-term success of current SUDs interventions are so low, new interventions are needed and NIBS should be thoroughly investigated. Specifically, these investigations should target downregulating reward processing and upregulating executive control for individuals with SUDs to combat the known imbalance in this disease (Bechara, 2005; Bickel et al., 2007).

ACKNOWLEDGMENTS

The authors are supported by the Intramural Research Program of the National Institute on Drug Abuse, National Institutes of Health, Baltimore, Maryland.

REFERENCES

Abou-Elseoud, A., Starck, T., Remes, J., Nikkinen, J., Tervonen, O., Kiviniemi, V., 2010. The effect of model order selection in group PICA. Hum. Brain Mapp. 31, 1207–1216. https://doi.org/10.1002/hbm.20929.

Achard, S., Bullmore, E., 2007. Efficiency and cost of economical brain functional networks. PLoS Comput. Biol. 3, e17, https://doi.org/10.1371/journal.pcbi.0030017.

Adinoff, B., Gu, H., Merrick, C., McHugh, M., Jeon-Slaughter, H., Lu, H., Yang, Y., Stein, E.A., 2015. Basal hippocampal activity and its functional connectivity predicts cocaine relapse. Biol. Psychiatry 78, 496–504. https://doi.org/10.1016/j.biopsych.2014.12.027.

Agam, Y., Hamalainen, M.S., Lee, A.K.C., Dyckman, K.A., Friedman, J.S., Isom, M., Makris, N., Manoach, D.S., 2011. Multimodal neuroimaging dissociates hemodynamic and electrophysiological correlates of error processing. Proc. Natl. Acad. Sci. 108, 17556–17561. https://doi.org/10.1073/pnas.1103475108.

Bechara, A., 2005. Decision making, impulse control and loss of willpower to resist drugs: a neurocognitive perspective. Nat. Neurosci. 8, 1458–1463. https://doi.org/10.1038/nn1584.

Berlim, M.T., van den Eynde, F., Tovar-Perdomo, S., Daskalakis, Z.J., 2014. Response, remission and drop-out rates following high-frequency repetitive transcranial magnetic stimulation (rTMS) for treating major depression: a systematic review and meta-analysis of randomized, double-blind and sham-controlled trials. Psychol. Med. 44, 225–239.

Bickel, W.K., Miller, M.L., Yi, R., Kowal, B.P., Lindquist, D.M., Pitcock, J.A., 2007. Behavioral and neuroeconomics of drug addiction: Competing neural systems and temporal discounting processes. Drug alcohol depend. Behav. Econ. Perspect. Drug Abuse Res. 90, S85–S91. https://doi.org/10.1016/j.drugalcdep.2006.09.016.

Breckel, T.P.K., Thiel, C.M., Giessing, C., 2013. The efficiency of functional brain networks does not differ between smokers and non-smokers. Psych. Res. Neuroimaging 214, 349–356. https://doi.org/10.1016/j.pscychresns.2013.07.005.

Calhoun, V.D., Liu, J., Adali, T., 2009. A review of group ICA for fMRI data and ICA for joint inference of imaging, genetic, and ERP data. Neuroimage 45, S163–S172. https://doi.org/10.1016/j.neuroimage.2008.10.057.

Camprodon, J.A., Martinez-Rega, J., Alonso-Alonso, M., Shih, M.-C., Pascual-Leone, A., 2007. One session of high frequency repetitive transcranial magnetic stimulation (rTMS) to the right prefrontal cortex transiently reduces cocaine craving. Drug Alcohol Depend. 86, 91–94. https://doi.org/10.1016/j.drugalcdep.2006.06.002.

Chanraud, S., Pitel, A.-L., Pfefferbaum, A., Sullivan, E.V., 2011. Disruption of functional connectivity of the default-mode network in alcoholism. Cereb. Cortex 21, 2272–2281. https://doi.org/10.1093/cercor/bhq297.

Chen, S., Ross, T.J., Zhan, W., Myers, C.S., Chuang, K.-S., Heishman, S.J., Stein, E.A., Yang, Y., 2008. Group independent component analysis reveals consistent resting-state networks across multiple sessions. Brain Res. 1239, 141–151. https://doi.org/10.1016/j.brainres.2008.08.028.

Cisler, J.M., Elton, A., Kennedy, A.P., Young, J., Smitherman, S., Andrew James, G., Kilts, C.D., 2013. Altered functional connectivity of the insular cortex across prefrontal networks in cocaine addiction. Psych. Res. Neuroimaging 213, 39–46. https://doi.org/10.1016/j.pscychresns.2013.02.007.

Cole, D.M., Beckmann, C.F., Long, C.J., Matthews, P.M., Durcan, M.J., Beaver, J.D., 2010. Nicotine replacement in abstinent smokers improves cognitive withdrawal symptoms with modulation of resting brain network dynamics. Neuroimage 52, 590–599. https://doi.org/10.1016/j.neuroimage.2010.04.251.

Cole, D.M., Oei, N.Y.L., Soeter, R.P., Both, S., van Gerven, J.M., Rombouts, S.A., Beckmann, C.F., 2013. Dopamine-dependent architecture of cortico-subcortical network connectivity. Cereb. Cortex 23, 1509–1516. https://doi.org/10.1093/cercor/bhs136.

Damoiseaux, J.S., Rombouts, S.A., Barkhof, F., Scheltens, P., Stam, C.J., Smith, S.M., Beckmann, C.F., 2006. Consistent resting-state networks across healthy subjects. Proc. Natl. Acad. Sci. 103, 13848–13853. https://doi.org/10.1073/pnas.0601417103.

Di Chiara, G., 2002. Nucleus accumbens shell and core dopamine: differential role in behavior and addiction. Behav. Brain Res. 137, 75–114. https://doi.org/10.1016/S0166-4328(02)00286-3.

Diana, M., Raij, T., Melis, M., Nummenmaa, A., Leggio, L., Bonci, A., 2017. Rehabilitating the addicted brain with transcranial magnetic stimulation. Nat. Rev. Neurosci. advance online publication https://doi.org/10.1038/nrn.2017.113.

Ding, X., Lee, S.-W., 2013. Changes of functional and effective connectivity in smoking replenishment on deprived heavy smokers: a resting-state fMRI study. PLoS One 8, e59331https://doi.org/10.1371/journal.pone.0059331.

Ding, X., Yang, Y., Stein, E.A., Ross, T.J., 2017. Combining multiple resting-state fMRI features during classification: optimized frameworks and their application to nicotine addiction. Front. Hum. Neurosci. 11, 362. https://doi.org/10.3389/fnhum.2017.00362.

Drysdale, A.T., Grosenick, L., Downar, J., Dunlop, K., Mansouri, F., Meng, Y., Fetcho, R.N., Zebley, B., Oathes, D.J., Etkin, A., Schatzberg, A.F., Sudheimer, K., Keller, J., Mayberg, H.S., Gunning, F.M., Alexopoulos, G.S., Fox, M.D., Pascual-Leone, A., Voss, H.U., Casey, B.J., Dubin, M.J., Liston, C., 2017. Resting-state connectivity biomarkers define neurophysiological subtypes of depression. Nat. Med. 23, 28–38. https://doi.org/10.1038/nm.4246.

Edwards, B.G., Calhoun, V.D., Kiehl, K.A., 2012. Joint ICA of ERP and fMRI during error-monitoring. Neuroimage 59, 1896–1903. https://doi.org/10.1016/j.neuroimage.2011.08.088.

Everitt, B.J., Robbins, T.W., 2005. Neural systems of reinforcement for drug addiction: from actions to habits to compulsion. Nat. Neurosci. 8, 1481–1489. https://doi.org/10.1038/nn1579.

Fedota, J.R., Matous, A.L., Salmeron, B.J., Gu, H., Ross, T.J., Stein, E.A., 2016. Insula demonstrates a non-linear response to varying demand for cognitive control and weaker resting connectivity with the executive control network in smokers. Neuropsychopharmacology 41, 2557–2565. https://doi.org/10.1038/npp.2016.62.

Fedota, J.R., Ding, X., Matous, A.L., Salmeron, B.J., McKenna, M.R., Gu, H., Ross, T.J., Stein, E.A., 2017. Nicotine abstinence influences the calculation of salience in discrete insular circuits. Biol. Psych. Cogn. Neurosci. Neuroimaging. https://doi.org/10.1016/j.bpsc.2017.09.010.

Feil, J., Zangen, A., 2010. Brain stimulation in the study and treatment of addiction. Neurosci. Biobehav. Rev. 34, 559–574. https://doi.org/10.1016/j.neubiorev.2009.11.006.

Filbey, F.M., Aslan, S., Calhoun, V.D., Spence, J.S., Damaraju, E., Caprihan, A., Segall, J., 2014. Long-term effects of marijuana use on the brain. Proc. Natl. Acad. Sci. U. S. A. 111, 16913–16918. https://doi.org/10.1073/pnas.1415297111.

Fink, B.C., Steele, V.R., Maurer, M.J., Fede, S.J., Calhoun, V.D., Kiehl, K.A., 2016. Brain potentials predict substance abuse treatment completion in a prison sample. Brain Behav. 6, https://doi.org/10.1002/brb3.501.

Forster, S.E., Carter, C.S., Cohen, J.D., Cho, R.Y., 2010. Parametric manipulation of the conflict signal and control-state adaptation. J. Cogn. Neurosci. 23, 923–935. https://doi.org/10.1162/jocn.2010.21458.

Fox, M.D., Snyder, A.Z., Vincent, J.L., Corbetta, M., Van Essen, D.C., Raichle, M.E., 2005. The human brain is intrinsically organized into dynamic, anticorrelated functional networks. Proc. Natl. Acad. Sci. U. S. A. 102, 9673–9678. https://doi.org/10.1073/pnas.0504136102.

Fox, M.D., Buckner, R.L., White, M.P., Greicius, M.D., Pascual-Leone, A., 2012a. Efficacy of transcranial magnetic stimulation targets for depression is related to intrinsic functional connectivity with the subgenual cingulate. Biol. Psychiatry 72, 595–603. https://doi.org/10.1016/j.biopsych.2012.04.028.

Fox, M.D., Halko, M.A., Eldaief, M.C., Pascual-Leone, A., 2012b. Measuring and manipulating brain connectivity with resting state functional connectivity magnetic resonance imaging (fcMRI) and transcranial magnetic stimulation (TMS). Neuroimage 62, 2232–2243. https://doi.org/10.1016/j.neuroimage.2012.03.035.

Garavan, H., Pankiewicz, J., Bloom, A., Cho, J.K., Sperry, L., Ross, T.J., Salmeron, B.J., Risinger, R., Kelley, D., Stein, E.A., 2000. Cue-induced cocaine craving: neuroanatomical specificity for drug users and drug stimuli. Am. J. Psychiatry 157, 1789–1798. https://doi.org/10.1176/appi.ajp.157.11.1789.

George, M.S., Wassermann, E.M., Williams, W.A., Callahan, A., Ketter, T.A., Basser, P., Hallett, M., Post, R.M., 1995. Daily repetitive transcranial magnetic stimulation (rTMS) improves mood in depression. Neuroreport 6, 1853–1856. https://doi.org/10.1097/00001756-199510020-00008.

Gießing, C., Thiel, C.M., 2012. Pro-cognitive drug effects modulate functional brain network organization. Front. Behav. Neurosci. 6, https://doi.org/10.3389/fnbeh.2012.00053.

Gießing, C., Thiel, C.M., Alexander-Bloch, A.F., Patel, A.X., Bullmore, E.T., 2013. Human brain functional network changes associated with enhanced and impaired attentional task performance. J. Neurosci. 33, 5903–5914. https://doi.org/10.1523/JNEUROSCI.4854-12.2013.

Goldstein, R.Z., Volkow, N.D., 2002. Drug addiction and its underlying neurobiological basis: neuroimaging evidence for the involvement of the frontal cortex. Am. J. Psychiatry 159, 1642–1652. https://doi.org/10.1176/appi.ajp.159.10.1642.

Gu, H., Salmeron, B.J., Ross, T.J., Geng, X., Zhan, W., Stein, E.A., Yang, Y., 2010. Mesocorticolimbic circuits are impaired in chronic cocaine users as demonstrated by resting-state functional connectivity. Neuroimage 53, 593–601. https://doi.org/10.1016/j.neuroimage.2010.06.066.

Guimerà, R., Amaral, L.A.N., 2005. Cartography of complex networks: modules and universal roles. J. Stat. Mech. P02001-1–P02001-13 https://doi.org/10.1088/1742-5468/2005/02/P02001.

Haber, S.N., Knutson, B., 2010. The reward circuit: linking primate anatomy and human imaging. Neuropsychopharmacology 35, 4–26. https://doi.org/10.1038/npp.2009.129.

Hanlon, C.A., Dowdle, L.T., Austelle, C.W., DeVries, W., Mithoefer, O., Badran, B.W., George, M.S., 2015. What goes up, can come down: novel brain stimulation paradigms may attenuate craving and craving-related neural circuitry in substance dependent individuals. Brain Res. 1628, 199–209. https://doi.org/10.1016/j.brainres.2015.02.053.

Hester, R., 2004. Executive dysfunction in cocaine addiction: evidence for discordant frontal, cingulate, and cerebellar activity. J. Neurosci. 24, 11017–11022. https://doi.org/10.1523/JNEUROSCI.3321-04.2004.

Hong, L.E., Gu, H., Yang, Y., Ross, T.J., Salmeron, B.J., Buchholz, B., Thaker, G.K., Stein, E.A., 2009. Association of nicotine addiction and nicotine's actions with separate cingulate cortex functional circuits. JAMA Psychiat. 66, 431–441. https://doi.org/10.1001/archgenpsychiatry.2009.2.

Hong, L.E., Hodgkinson, C.A., Yang, Y., Sampath, H., Ross, T.J., Buchholz, B., Salmeron, B.J., Srivastava, V., Thaker, G.K., Goldman, D., Stein, E.A., 2010. A genetically modulated, intrinsic cingulate circuit supports human nicotine addiction. Proc. Natl. Acad. Sci. U. S. A. 107, 13509–13514. https://doi.org/10.1073/pnas.1004745107.

Hu, Y., Salmeron, B.J., Gu, H., Stein, E.A., Yang, Y., 2015. Impaired functional connectivity within and between Frontostriatal circuits and its association with compulsive drug use and trait impulsivity in cocaine addiction. JAMA Psychiat. 72, 584–592. https://doi.org/10.1001/jamapsychiatry.2015.1.

Jansen, J.M., Daams, J.G., Koeter, M.W.J., Veltman, D.J., van den Brink, W., Goudriaan, A.E., 2013. Effects of non-invasive neurostimulation on craving: a meta-analysis. Neurosci. Biobehav. Rev. 37, 2472–2480. https://doi.org/10.1016/j.neubiorev.2013.07.009.

Jiang, G., Wen, X., Qiu, Y., Zhang, R., Wang, J., Li, M., Ma, X., Tian, J., Huang, R., 2013. Disrupted topological organization in whole-brain functional networks of heroin-dependent individuals: a resting-state fMRI study. PLoS One 8, e82715, https://doi.org/10.1371/journal.pone.0082715.

Kalivas, P.W., O'Brien, C., 2008. Drug addiction as a pathology of staged neuroplasticity. Neuropsychopharmacology 33, 166. https://doi.org/10.1038/sj.npp.1301564.

Kaufman, J.N., Ross, T.J., Stein, E.A., Garavan, H., 2003. Cingulate hypoactivity in cocaine users during a Go-Nogo task as revealed by event-related functional magnetic resonance imaging. J. Neurosci. 23, 7839–7843. https://doi.org/10.1523/JNEUROSCI.23-21-07839.2003.

Keck, M.E., Welt, T., Muller, M.B., Erhardt, A., Ohl, H., Toschi, N., Holsboer, F., Sillaber, I., 2002. Repetitive transcranial magnetic stimulation increases the release of dopamine in the mesolimbic and meostriatal system. Neuropharmacology 43, 101–109. https://doi.org/10.1016/S0028-3908(02)00069-2.

Kelley, A.E., 2004. Memory and addiction: shared neural circuitry and molecular mechanisms. Neuron 44, 161–179. https://doi.org/10.1016/j.neuron.2004.09.016.

Kelly, C., Zuo, X.N., Gotimer, K., Cox, C.L., Lynch, L., Brock, D., Imperati, D., Garavan, H., Rotrosen, J., Castellanos, F.X., Milham, M.P., 2011. Reduced interhemispheric resting state functional connectivity in cocaine addiction. Biol. Psychiatry 69, 684–692. https://doi.org/10.1016/j.biopsych.2010.11.022.

Konova, A.B., Moeller, S.J., Tomasi, D., Volkow, N.D., Goldstein, R.Z., 2013. Effects of methylphenidate on resting-state functional connectivity of the Mesocorticolimbic dopamine pathways in cocaine addiction. JAMA Psychiat. 70, 857–868. https://doi.org/10.1001/jamapsychiatry.2013.1129.

Koob, G.F., Volkow, N.D., 2016. Neurobiology of addiction: a neurocircuitry analysis. Lancet Psych. 3, 760–773. https://doi.org/10.1016/s2215-0366(16)00104-8.

Kühn, S., Gallinat, J., 2011. Common biology of craving across legal and illegal drugs - a quantitative meta-analysis of cue-reactivity brain response: common biology of craving across legal and illegal drugs. Eur. J. Neurosci. 33, 1318–1326. https://doi.org/10.1111/j.1460-9568.2010.07590.x.

Lerman, C., Gu, H., Loughead, J., Ruparel, K., Yang, Y., Stein, E.A., 2014. Large-scale brain network coupling predicts acute nicotine abstinence effects on craving and cognitive function. JAMA Psychiat. 71, 523–530. https://doi.org/10.1001/jamapsychiatry.2013.4091.

Li, Y.-O., Adalı, T., Calhoun, V.D., 2007. Estimating the number of independent components for functional magnetic resonance imaging data. Hum. Brain Mapp. 28, 1251–1266. https://doi.org/10.1002/hbm.20359.

Li, X., Hartwell, K.J., Owens, M., Lematty, T., Borckardt, J.J., Hanlon, C.A., Brady, K.T., George, M.S., 2013. Repetitive transcranial magnetic stimulation of the dorsolateral prefrontal cortex reduces nicotine cue craving. Biol. Psychiatry 73, 714–720. https://doi.org/10.1016/j.biopsych.2013.01.003.

Li, W., Li, Q., Wang, D., Xiao, W., Liu, K., Shi, L., Zhu, J., Li, Y., Yan, X., Chen, J., Ye, J., Li, Z., Wang, Y., Wang, W., 2015. Dysfunctional default mode network in methadone treated patients who have a higher heroin relapse risk. Sci. Rep. 5, srep15181. https://doi.org/10.1038/srep15181.

Liang, X., He, Y., Salmeron, B.J., Gu, H., Stein, E.A., Yang, Y., 2015. Interactions between the salience and default-mode networks are disrupted in cocaine addiction. J. Neurosci. 35, 8081–8090. https://doi.org/10.1523/JNEUROSCI.3188-14.2015.

Lin, F., Wu, G., Zhu, L., Lei, H., 2015. Altered brain functional networks in heavy smokers. Addict. Biol. 20, 809–819. https://doi.org/10.1111/adb.12155.

Liu, J., Liang, J., Qin, W., Tian, J., Yuan, K., Bai, L., Zhang, Y., Wang, W., Wang, Y., Li, Q., Zhao, L., Lu, L., von Deneen, K.M., Liu, Y., Gold, M.S., 2009. Dysfunctional connectivity patterns in chronic heroin users: an fMRI study. Neurosci. Lett. 460, 72–77. https://doi.org/10.1016/j.neulet.2009.05.038.

Liu, J., Qin, W., Yuan, K., Li, J., Wang, W., Li, Q., Wang, Y., Sun, J., von Deneen, K.M., Liu, Y., Tian, J., 2011. Interaction between dysfunctional connectivity at rest and heroin cues-induced brain responses in male abstinent heroin-dependent individuals. PLoS One 6, e23098, https://doi.org/10.1371/journal.pone.0023098.

Ma, N., Liu, Y., Li, N., Wang, C.X., Zhang, H., Jiang, X.F., Xu, H.S., Fu, X.M., Hu, X., Zhang, D.R., 2010. Addiction related alteration in resting-state brain connectivity. Neuroimage 49, 738–744. https://doi.org/10.1016/j.neuroimage.2009.08.037.

Ma, N., Liu, Y., Fu, X.M., Li, N., Wang, C.X., Zhang, H., Qian, R.B., Xu, H.S., Hu, X., Zhang, D.R., 2011. Abnormal brain default-mode network functional connectivity in drug addicts. PLoS One 6, e16560, https://doi.org/10.1371/journal.pone.0016560.

Ma, X., Qiu, Y., Tian, J., Wang, J., Li, S., Zhan, W., Wang, T., Zeng, S., Jiang, G., Xu, Y., 2015. Aberrant default-mode functional and structural connectivity in heroin-dependent individuals. PLoS One 10, e0120861. https://doi.org/10.1371/journal.pone.0120861.

Marhe, R., van de Wetering, B.J.M., Franken, I.H.A., 2013. Error-related brain activity predicts cocaine use after treatment at 3-month follow-up. Biol. Psychiatry 73, 782–788. https://doi.org/10.1016/j.biopsych.2012.12.016.

McHugh, M.J., Demers, C.H., Braud, J., Briggs, R., Adinoff, B., Stein, E.A., 2013. Striatal-insula circuits in cocaine addiction: implications for impulsivity and relapse risk. Am. J. Drug Alcohol Abuse 39, 424–432. https://doi.org/10.3109/00952990.2013.847446.

McHugh, M.J., Demers, C.H., Salmeron, B.J., Devous Sr., M.D., Stein, E.A., Adinoff, B., 2014. Cortico-amygdala coupling as a marker of early relapse risk in cocaine-addicted individuals. Front. Psych. 5, 16. https://doi.org/10.3389/fpsyt.2014.00016.

McHugh, M.J., Gu, H., Yang, Y., Adinoff, B., Stein, E.A., 2016. Executive control network connectivity strength protects against relapse to cocaine use. Addict. Biol. https://doi.org/10.1111/adb.12448.

Mckeown, M.J., Makeig, S., Brown, G.G., Jung, T.-P., Kindermann, S.S., Bell, A.J., Sejnowski, T.J., 1998. Analysis of fMRI data by blind separation into independent spatial components. Hum. Brain Mapp. 6, 160–188. https://doi.org/10.1002/(SICI)1097-0193(1998)6:3<160::AID-HBM5>3.0.CO;2-1.

McKiernan, K.A., Kaufman, J.N., Kucera-Thompson, J., Binder, J.R., 2003. A parametric manipulation of factors affecting task-induced deactivation in functional neuroimaging. J. Cogn. Neurosci. 15, 394–408. https://doi.org/10.1162/089892903321593117.

Menon, V., Uddin, L.Q., 2010. Saliency, switching, attention and control: a network model of insula function. Brain Struct. Funct. 214, 655–667. https://doi.org/10.1007/s00429-010-0262-0.

Mishra, B.R., Nizamie, S.H., Das, B., Praharaj, S.K., 2010. Efficacy of repetitive transcranial magnetic stimulation in alcohol dependence: a sham-controlled study. Addiction 105, 49–55. https://doi.org/10.1111/j.1360-0443.2009.02777.x.

Naqvi, N.H., Bechara, A., 2009. The hidden island of addiction: the insula. Trends Neurosci. 32, 56–67. https://doi.org/10.1016/j.tins.2008.09.009.

Naqvi, N.H., Rudrauf, D., Damasio, H., Bechara, A., 2007. Damage to the insula disrupts addiction to cigarette smoking. Science 315, 531–534. https://doi.org/10.1126/science.1135926.

Nestler, E.J., 2005. Is there a common molecular pathway for addiction? Nat. Neurosci. 8, 1445–1449. https://doi.org/10.1038/nn1578.

Newman, M.E.J., 2006. Modularity and community structure in networks. Proc. Natl. Acad. Sci. 103, 8577–8582. https://doi.org/10.1073/pnas.0601602103.

Nutt, D.J., Lingford-Hughes, A., Erritzoe, D., Stokes, P.R.A., 2015. The dopamine theory of addiction: 40 years of highs and lows. Nat. Rev. Neurosci. 16, 305–312. https://doi.org/10.1038/nrn3939.

Pariyadath, V., Stein, E.A., Ross, T.J., 2014. Machine learning classification of resting state functional connectivity predicts smoking status. Front. Hum. Neurosci. 8, 425. https://doi.org/10.3389/fnhum.2014.00425.

Pascual-leone, A., Rubio, B., Pallardó, F., Catalá, M.D., 1996. Rapid-rate transcranial magnetic stimulation of left dorsolateral prefrontal cortex in drug-resistant depression. Lancet 348, 233. https://doi.org/10.1016/S0140-6736(96)01219-6.

Petri, G., Expert, P., Turkheimer, F., Carhart-Harris, R., Nutt, D., Hellyer, P.J., Vaccarino, F., 2014. Homological scaffolds of brain functional networks. J. R. Soc. Interface 11, 20140873. https://doi.org/10.1098/rsif.2014.0873.

Politi, E., Fauci, E., Santoro, A., Smeraldi, E., 2008. Daily sessions of transcranial magnetic stimulation to the left prefrontal cortex gradually reduce cocaine craving. Am. J. Addict. 17, 345–346. https://doi.org/10.1080/10550490802139283.

Raichle, M.E., 2015. The brain's default mode network. Annu. Rev. Neurosci. 38, 433–447. https://doi.org/10.1146/annurev-neuro-071013-014030.

Sahlem, G.L., Breedlove, J., Taylor, J.J., Badran, B.A., Lauer, A., George, M.S., Brady, K.T., Borckardt, J.J., Back, S.E., Hanlon, C.A., 2017. Dorsolateral prefrontal cortex transcranial magnetic stimulation as a tool to decrease pain and craving in opiate dependent individuals: a pilot study of feasibility and effect size. Brain Stimul. Basic Transl. Clin. Res. Neuromod. 10, 482. https://doi.org/10.1016/j.brs.2017.01.412.

Schlam, T.R., Baker, T.B., 2013. Interventions for tobacco smoking. Annu. Rev. Clin. Psychol. 9, 675–702. https://doi.org/10.1146/annurev-clinpsy-050212-185602.

Schmidt, A., Denier, N., Magon, S., Radue, E.-W., Huber, C.G., Riecher-Rossler, A., Wiesbeck, G.A., Lang, U.E., Borgwardt, S., Walter, M., 2015. Increased functional connectivity in the resting-state basal ganglia network after acute heroin substitution. Transl. Psychiatry 5, e533, https://doi.org/10.1038/tp.2015.28.

Seeley, W.W., Menon, V., Schatzberg, A.F., Keller, J., Glover, G.H., Kenna, H., Reiss, A.L., Greicius, M.D., 2007. Dissociable intrinsic connectivity networks for salience processing and executive control. J. Neurosci. 27, 2349–2356. https://doi.org/10.1523/JNEUROSCI.5587-06.2007.

Sjoerds, Z., Stufflebeam, S.M., Veltman, D.J., Van den Brink, W., Penninx, B.W.J.H., Douw, L., 2017. Loss of brain graph network efficiency in alcohol dependence. Addict. Biol. 22, 523–534. https://doi.org/10.1111/adb.12346.

Smith, S.M., Fox, P.T., Miller, K.L., Glahn, D.C., Fox, P.M., Mackay, C.E., Filippini, N., Watkins, K.E., Toro, R., Laird, A.R., Beckmann, C.F., 2009. Correspondence of the brain's functional architecture during activation and rest. Proc. Natl. Acad. Sci. U. S. A. 106, 13040–13045. https://doi.org/10.1073/pnas.0905267106.

Spronk, D.B., van Wel, J.H.P., Ramaekers, J.G., Verkes, R.J., 2013. Characterizing the cognitive effects of cocaine: a comprehensive review. Neurosci. Biobehav. Rev. 37, 1838–1859. https://doi.org/10.1016/j.neubiorev.2013.07.003.

Sridharan, D., Levitin, D.J., Menon, V., 2008. A critical role for the right fronto-insular cortex in switching between central-executive and default-mode networks. Proc. Natl. Acad. Sci. 105, 12569–12574. https://doi.org/10.1073/pnas.0800005105.

Steele, V.R., Fink, B.C., Maurer, J.M., Arbabshirani, M.R., Wilber, C.H., Jaffe, A.J., Sidz, A., Pearlson, G.D., Calhoun, V.D., Clark, V.P., Kiehl, K.A., 2014. Brain potentials measured during a go/NoGo task predict completion of substance abuse treatment. Biol. Psychiatry 76, 75–83. https://doi.org/10.1016/j.biopsych.2013.09.030.

Steele, V.R., Pariyadath, V., Goldstein, R.Z., Stein, E.A., 2017. Reward cirtuitry and drug addiciton. In: Charney, D.S., Nestler, E.J., Buxbaum, J., Sklar, P. (Eds.), Neurobiology of Mental Illness. fifth ed. Oxford University Press, Oxford.

Steele, V.R., Maurer, J.M., Arbabshirani, M.R., Claus, E.D., Fink, B.C., Rao, V., Calhoun, V.D., Kiehl, K.A., 2018. Machine learning of functional magnetic resonance imaging network connectivity predicts substance abuse treatment completion. Biol. Psych. Cogn. Neurosci. Neuroimaging. https://doi.org/10.1016/j.bpsc.2017.07.003.

Strafella, A.P., Paus, T., Barrett, J., Dagher, A., 2001. Repetitive transcranial magnetic stimulation of the human prefrontal cortex induces dopamine release in the caudate nucleus. J. Neurosci. 21, 1–4.

Sutherland, M.T., McHugh, M.J., Pariyadath, V., Stein, E.A., 2012. Resting state functional connectivity in addiction: lessons learned and a road ahead. Neuroimage 62, 2281–2295. https://doi.org/10.1016/j.neuroimage.2012.01.117.

Sutherland, M.T., Carroll, A.J., Salmeron, B.J., Ross, T.J., Hong, L.E., Stein, E.A., 2013a. Down-regulation of amygdala and insula functional circuits by varenicline and nicotine in abstinent cigarette smokers. Biol. Psychiatry 74, 538–546. https://doi.org/10.1016/j.biopsych.2013.01.035.

Sutherland, M.T., Carroll, A.J., Salmeron, B.J., Ross, T.J., Stein, E.A., 2013b. Insula's functional connectivity with ventromedial prefrontal cortex mediates the impact of trait alexithymia on state tobacco craving. Psychopharmacology (Berl) 228, 143–155. https://doi.org/10.1007/s00213-013-3018-8.

Tanabe, J., Nyberg, E., Martin, L.F., Martin, J., Cordes, D., Kronberg, E., Tregellas, J.R., 2011. Nicotine effects on default mode network during resting state. Psychopharmacology (Berl) 216, 287–295. https://doi.org/10.1007/s00213-011-2221-8.

Terraneo, A., Leggio, L., Saladini, M., Ermani, M., Bonci, A., Gallimberti, L., 2016. Transcranial magnetic stimulation of dorsolateral prefrontal cortex reduces cocaine use: a pilot study. Eur. Neuropsychopharmacol. 26, 37–44. https://doi.org/10.1016/j.euroneuro.2015.11.011.

Volkow, N.D., Morales, M., 2015. The brain on drugs: From reward to addiction. Cell 162, 712–725. https://doi.org/10.1016/j.cell.2015.07.046.

Volkow, N.D., Koob, G.F., McLellan, A.T., 2016. Neurobiologic advances from the brain disease model of addiction. N. Engl. J. Med. 374, 363–371. https://doi.org/10.1056/NEJMra1511480.

Wang, Z., Suh, J., Li, Z., Li, Y., Franklin, T., O'Brien, C., Childress, A.R., 2015. A hyperconnected but less efficient small-world network in the substance-dependent brain. Drug Alcohol Depend. 152, 102–108. https://doi.org/10.1016/j.drugalcdep.2015.04.015.

Wang, P.-W., Lin, H.-C., Liu, G.-C., Yang, Y.-H.C., Ko, C.-H., Yen, C.-F., 2016. Abnormal interhemispheric resting state functional connectivity of the insula in heroin users under methadone maintenance treatment. Psych. Res. Neuroimaging 255, 9–14. https://doi.org/10.1016/j.pscychresns.2016.07.009.

Watts, D.J., Strogatz, S.H., 1998. Collective dynamics of 'small-world' networks. Nature 393, 440–442. https://doi.org/10.1038/30918.

Wetherill, R.R., Fang, Z., Jagannathan, K., Childress, A.R., Rao, H., Franklin, T.R., 2015. Cannabis, cigarettes, and their co-occurring use: disentangling differences in default mode network functional connectivity. Drug Alcohol Depend. 153, 116–123. https://doi.org/10.1016/j.drugalcdep.2015.05.046.

Wise, R.A., 2002. Brain reward circuitry: insights from unsensed incentives. Neuron 36, 229–240. https://doi.org/10.1016/S0896-6273(02)00965-0.

Worhunsky, P.D., Stevens, M.C., Carroll, K.M., Rounsaville, B.J., Calhoun, V.D., Pearlson, G.D., Potenza, M.N., 2013. Functional brain networks associated with cognitive control, cocaine dependence, and treatment outcome. Psychol. Addict. Behav. 27, 477–488. https://doi.org/10.1037/a0029092.

Yuan, K., Qin, W., Liu, J., Guo, Q., Dong, M., Sun, J., Zhang, Y., Liu, P., Wang, W., Wang, Y., Li, Q., Yang, W., von Deneen, K.M., Gold, M.S., Liu, Y., Tian, J., 2010. Altered small-world brain functional networks and duration of heroin use in male abstinent heroin-dependent individuals. Neurosci. Lett. 477, 37–42. https://doi.org/10.1016/j.neulet.2010.04.032.

Connectivity and dysconnectivity: A brief history of functional connectivity research in schizophrenia and future directions

Eva Mennigen*, Barnaly Rashid[†], Vince D. Calhoun[‡,§]

Department of Psychiatry and Biobehavioral Sciences, Semel Institute for Neuroscience and Human Behavior, University of California, Los Angeles, CA, United States* Department of Psychiatry, Harvard Medical School, Boston, MA, United States[†] The Mind Research Network, Albuquerque, NM, United States[‡] Department of Electrical and Computer Engineering, University of New Mexico, Albuquerque, NM, United States[§]

CHAPTER OUTLINE

INTRODUCTION

Unfortunately, in the field of psychic disturbances there is not a single symptom which is pathognomic for any particular illness. On the other hand, we can expect that the composition of the individual characteristics which form the total picture, and in particular the changes which develop in the course of the illness, will not be produced in exactly the same way by any of the other diseases...

(Kraepelin, 1899) (English translation (Hoenig, 1983))

Connectomics. https://doi.org/10.1016/B978-0-12-813838-0.00007-8

With the first steps toward establishing modern psychiatry, and the first attempts to classify different mental phenomena and disorders into clearly defined clusters, the need arose for diagnostic tools to allow for an objective evaluation of symptoms. After one century into this experiment and the idea of classifying psychiatric disorders based on predefined symptom categories, we are currently at a turning point where a new paradigm has emerged: The Research Domain Criteria (RDoC; Insel et al., 2010). This approach aims at incorporating the most recent findings from clinical and genetic neuroscience, thereby opening the field to a dimensional approach informed by the specific neural pathophysiology underlying psychiatric disorders. The information gained from relevant, reliable, and up-to-date neuroscientific research should—in the best case—educate patients and advise clinicians to guide them toward a personalized treatment, including prognosis evaluation and selection of the most appropriate psychopharmacological and psychotherapeutic treatments for individual patients. Tools to assess abnormal brain physiology to achieve this ambitious goal encompass investigations on micro- as well as macroscopic levels, ranging from genetic assessments (O'Donovan, 2015), to cell recordings (e.g., in epilepsy [Fried et al., 2014; Merricks et al., 2015]), diverse imaging techniques, foremost positron emission tomography (PET, e.g., in schizophrenia [Di Biase et al., 2017; Poels et al., 2014; Weinstein et al., 2017]), electro-encephalography (EEG, e.g., in schizophrenia [Khanna et al., 2015; Senkowski and Gallinat, 2015]), and magnetic resonance imaging (MRI, which will be reviewed in more depth later); each of these allows for a different degree of spatial and temporal resolution. A special focus with regard to MRI research lies on the examination of connectivity: the analysis of how different brain areas "talk" to each other. The brain's organization of integration vs. segregation implies that connectivity plays a pivotal role in this composition given the cognitive flexibility observed in mammals. Segregation refers to the fact that the brain is equipped with specialized functional areas that are spatially coherent. For example, the primary auditory cortex as one functional entity is located in one coherent brain area, the superior temporal gyrus. But these highly specialized areas, taken by themselves, would not allow for higher cognitive functions as seen in the mammalian brain. Only the integration of information from different highly specialized brain areas renders adaptation to environmental changes possible by dynamic network interactions (Cohen and D'Esposito, 2016; Deco et al., 2015). This theoretical framework has fueled the field's interest in connectivity analysis with a special focus on neuropsychiatric disorders, because these often exhibit a wide variety of symptoms ranging from emotional, cognitive, and behavioral alterations to thought disorders, and therefore imply dysconnectivity between different functional networks.

SCHIZOPHRENIA

Schizophrenia is not only one of the most common psychiatric disorders (the most recent epidemiologic review stated a median lifetime prevalence of 0.48% ([Simeone et al., 2015]), but it is also one of the most debilitating diseases with increased mortality (Laursen et al., 2014; Nielsen et al., 2013; Olfson et al., 2015), increased suicide risk (Hawton and van Heeringen, 2009; Hor and Taylor, 2010; Kelleher et al., 2013), and reduced quality of life (Karow et al., 2014). Clinically, schizophrenia is diagnosed based on characteristic symptoms such as hallucinations; delusions; disorganized thinking and speech; abnormal psychomotor behavior; and negative symptoms such as apathy, blunted affect, and anhedonia; as well as duration of symptoms (> 6 months). Moreover, schizophrenia is also accompanied by cognitive impairments and a decrease in social and occupational functioning, often leading to unemployment, homelessness, and stigmatization of patients suffering from schizophrenia (Folsom and Jeste, 2002). An insightful first-person report (Yeiser, 2017) and an analysis on the portrayal of schizophrenia by entertainment media (Owen, 2012) offer additional data.

The origin of the term *schizophrenia* is complex and majorly influenced by Emil Kraepelin (1865–1926) and Eugen Bleuer (1857–1939); Kraeplin paved the way for a modern diagnostic construct in psychiatry and first introduced the distinction of mood disorders vs psychosis. With regard to schizophrenia, he stuck with the term *dementia praecox,* which was coined in 1891 by Arnold Pick (1851–1924). In 1908, Bleuler introduced the term *schizophrenia* because longitudinal patient data showed that dementia was not a necessary part of the disorder itself and that the course of the disease is more variable than previously assumed. According to Bleuler (1908): "…*In my opinion the breaking up or splitting of psychic functioning is an excellent symptom of the whole group* (n.b., "the whole group" refers to the different subtypes of schizophrenia identified by Bleuler)". In this sentence, Bleuler explains his decision to call this condition *schizophrenia,* which translates to "splitting of the mind" from its Greek roots. That the concept and naming of schizophrenia might be overall outdated is the subject of an ongoing scholarly debate (Sato, 2006; Van Os, 2009).

The first-line treatment choice nowadays is a pharmacological intervention with antipsychotic drugs, which are dopamine receptor 2 (D2) antagonists with slightly different receptor profiles (Kusumi et al., 2015; Lally and MacCabe, 2015). Unfortunately, there were no significant discoveries with regard to more potent and fewer side-effect-laden agents in the past decades, emphasizing the need to understand more about the actual brain pathology underlying schizophrenia to guide development of new medications (Keshavan et al., 2017). The effectiveness of D2 antagonists has given rise to the hypothesis that positive symptoms in schizophrenia are caused by an excess of dopamine in mesolimbic brain areas (Grace, 1991, 2012), an idea supported by the observation that D2 agonists, on the other hand, can cause psychotic symptoms, as often occurs in the dopamine-substituting treatment of Parkinson's disease, for example (Ecker et al., 2009). Nevertheless, the dopamine hypothesis of schizophrenia falls short when trying to explain the diversity of negative and cognitive symptoms. The observation that certain anesthetics, namely ketamine, can mimic not only positive but also negative and cognitive symptoms of schizophrenia has led to the formulation of the *N*-methyl-D-aspartate-receptor (NMDAR) hypothesis, which involves the neurotransmitter glutamate as the main binding agent (Balu, 2016; Coyle, 2012). Current genetic studies further emphasize the involvement of glutamatergic synapses and NMDARs in the etiology of schizophrenia (Hall et al., 2015). Despite the enormous efforts in understanding the neurobiology of this disease, a large part of the disease pathology remains unknown (Franke et al., 2016; The Network and Pathway Analysis Subgroup of the Psychiatric Genomics Consortium, 2015). Besides the high heritability and genetic aspects of schizophrenia vulnerability, environmental factors (Shorter and Miller, 2015) and perinatal stress such as maternal viral infections and other obstetric complications play an important role in schizophrenia development (Cannon et al., 2000; Schmitt et al., 2014).

In summary, nowadays schizophrenia is viewed as a neurodevelopmental disorder with a high genetic liability that might be triggered by (early-)life stressors. However, the first signs and symptoms of schizophrenia do not usually emerge until adolescence, emphasizing this as a period of particular importance. Also, the onset of schizophrenia is usually not abrupt but is often proceeded by a phase retrospectively called prodrome that is typically characterized by attenuated psychotic symptoms and a decline in social and occupational functioning prior to meeting diagnostic criteria of schizophrenia (Carpenter and Schiffman, 2015; McGlashan et al., 2010).

The different neuronal domains altered in schizophrenia, as implied by the diverse clinical symptoms, in addition to the temporal trajectory of schizophrenia heavily imply that impaired brain connectivity might be the first and foremost disease correlate.

CONNECTIVITY

The brain itself is conceptualized as an intricate mesh with complex organizational principles enabling a concerted communication between single neurons, distinct neuron populations, and, in the end, remote brain areas. This communication is the center of many investigations aimed at elucidating neural pathophysiology of psychiatric disorders. In terms of MRI, evidence of communication is often called connectivity. Connectivity can be investigated at many levels. However, the most widely used and relatively easy accessible tools (for the most part) in studying human neural connectivity in vivo are MRI-based approaches. Given the anatomical differentiation of the brain into gray matter (GM) and white matter (WM), we can discern two major streams in MRI connectivity research: structural and functional connectivity.

STRUCTURAL CONNECTIVITY

Connectivity based on structural imaging is concerned with WM fiber tracts, their integrity, and thickness. For example, structural connectivity can be investigated by means of diffusion MRI (dMRI [Jones and Leemans, 2011; Soares et al., 2013]). Briefly, dMRI examines movement of water molecules, which is thought to be faster and less hindered in a direction along axons/fiber tracts rather than orthogonal to them. Metrics most commonly derived from dMRI through diffusion tensor approaches are fractional anisotropy (FA) and mean diffusivity (MD); both reflect different structural properties and are meant to be indicative of WM microstructure assembly.

MD measures how freely water molecules can diffuse in any direction, whereas FA is thought to reflect those constrains of water diffusion. Together, they can be used to describe the brain's structural integrity. In terms of schizophrenia research, it has overwhelmingly been shown that patients exhibit decreased FA and increased MD, which appears to be a global observation, traceable in unaffected siblings and adolescent offspring of patients with schizophrenia, and the effect appears to decrease with age (Canu et al., 2015; de Leeuw et al., 2015, 2017; Filippi et al., 2014; Kanaan et al., 2017; Samartzis et al., 2014). Moreover, by following the diffusion metrics from voxel to voxel, the three-dimensional fiber orientation can be reconstructed and paths of fast diffusion can be visualized, believed to represent underlying fiber tracts and axonal architecture. This method is called tractography (Hagmann et al., 2006; Mori and van Zijl, 2002). With regard to schizophrenia, this method showed that, in particular, fiber tracts connecting the frontal cortex and limbic brain areas (i.e., the uncinate fasciculus) and tracts between frontal and parietal cortices (i.e., the arcuate fasciculus) are disrupted (Karlsgodt et al., 2008; Voineskos et al., 2010; Wu et al., 2015).

Generally, structural connectivity reflects the physical backbone of functional measures of connectivity (Honey et al., 2009; Koch et al., 2002; Skudlarski et al., 2008). However, agreement between structural connectivity and functional connectivity is particularly high in densely connected brain areas and less so in other areas, indicating that both techniques are qualified to reveal different aspects of connectivity.

FUNCTIONAL CONNECTIVITY

Functional connectivity (FC), in contrast to structural connectivity, is based on the assumption that we can record changes of the blood oxygenation-level dependent (BOLD) signal, which is assumed to originate in the brain's GM due to the changes in neuronal activity. Most often, FC is estimated based on resting-state fMRI, which simply records the spontaneously occurring fluctuations in the BOLD signal while the study participant lies still in the MRI scanner either with their eyes open or closed, instructed to stay awake and to let their minds "wander" freely. FC refers to the temporal covariance/correlation of the BOLD time series of two or more spatially distinct brain areas, and it completes the picture of brain connectivity in addition to structural connectivity. Resting-state fMRI contrasts task-based fMRI where stimuli presented to the study participant while scanning are purposefully manipulated to investigate specific psychological phenomena (e.g., working memory, decision making, etc.). In the analysis of task-based fMRI, events are fit to the time courses of the BOLD signal after convolving the time course with a canonical hemodynamic response function. Afterward, the effects on the neuronal-originated BOLD response of specific manipulations can be estimated and conclusions can be drawn from these observations. In addition to being used to examine task-induced brain activity, task-based fMRI also lends itself to FC analyses. We will provide remarks on FC studies derived from both resting-state and task-based fMRI.

The first observation of FC during resting-state fMRI was made by Biswal et al. (1995) in the motor cortex, where they correlated time courses of regions identified by a finger-tapping task at rest and found a high correlation between bilateral motor regions. Nevertheless, it took a few years to convince the neuroimaging community that the signals measured at rest are indeed meaningful and relevant, and not fully attributable to noise (Fox and Raichle, 2007).

Another pivotal discovery of resting-state imaging was the observation of the so-called default mode network (DMN): a network of medial anterior as well as medial and lateral posterior brain areas showing increased activity at rest and decreased activity during states of task engagement (Raichle et al., 2001). (For perspectives challenging this simplified view, see Shine et al., 2016; Vatansever et al., 2015.) Because of this hypothetical diametric activation of the DMN and brain areas involved in task processing (generally termed task-positive network, TPN), researchers became interested in exploring DMN connectivity in more detail, particularly in diseases associated with cognitive impairments as it seemed likely that dysconnectivity between these networks and cognitive impairment might be associated. Since then, a wealth of literature has been published aimed at examining the functional role of the DMN as well as its involvement in disease (Raichle, 2015; Whitfield-Gabrieli and Ford, 2012).

With regard to schizophrenia, another important piece in the puzzle was an early task-based PET study by Friston and Frith (1995). Here, the authors not only revisit Bleuler's idea of characterizing schizophrenia as a disorder of disintegrated brain functions (Bleuler, 1908, 1911), but in fact they demonstrate cortical dysconnectivity between distinct frontotemporal brain networks in patients with schizophrenia. Further, they stress that aberrant integration of perception might cause positive symptoms, a hypothesis that has been further elaborated and found expressed in a computational model of psychosis (Adams et al., 2013).

A lot of insight has been gained over the years, and novel methodological approaches are emerging constantly to refine our understanding of dysconnectivity in schizophrenia. In terms of FC, we can classify three broad approaches that are being used: graph theoretical approaches (based on network sciences), seed-based approaches or region of interest (ROI) analyses, and independent component analysis (ICA, i.e., data-driven approach).

Graph theoretic approaches use parcellation algorithms (which can include ICA and ROI approaches) to first segment the brain's GM into smaller portions before the actual connectivity analysis. The other two approaches investigate temporal covariation of time courses either from predefined ROIs or seeds[1] or from ICA derived from the

---

[1] N.b., the difference between both seems arbitrary, nevertheless some authors argue that ROI refers to a bigger brain area or network, whereas a seed is usually one particular anatomical landmark of some chosen volume; both can be derived either from anatomical atlases or prior studies.

whole-brain-spanning intrinsic connectivity networks (ICNs). We will present detailed discussion on each approach, its strengths and weaknesses, and findings with regard to schizophrenia yielded using the particular approach in the following paragraphs.

GRAPH THEORY
Technical foundations

As a branch of network science, graph theory has recently been adapted to explore FC in the human brain. Within this framework, neurons or brain regions are called "nodes", whereas "edges" reflect the connectivity between these nodes. A graph represents the entire network of the brain including the interconnections between the nodes. The steps involved in order to yield a graph theoretical representation of brain connectivity are as follows: preprocessing of rsfMRI data including thorough processes to eliminate subjects' motion (Power et al., 2014); parcellation of brain regions, for example, based on anatomical atlases (Tzourio-Mazoyer et al., 2002), functional mapping, or data-driven approaches; estimation of covariance (correlation or coherence) between all identified brain regions yielding a connectivity matrix; then binarizing the adjacency matrix by applying an arbitrary threshold or using a weighted approach yielding a directed graph representation. Finally, connections surviving this threshold are converted into a graph representation, and the organization of this graph can be further investigated; steps are visualized in Fig. 1.

Based on nodes and edges, a variety of metrics can be estimated characterizing the graph. Some of the most widely used metrics with regard to describing network characteristics are:

- Clustering: the tendency of nodes to form local clusters (i.e., densely interconnected nodes)
- Paths length: a distance measure (i.e., how many other nodes have to be passed when traveling from node A to B)
- Average shortest path length: also known as characteristic path length (i.e., the average number of nodes that have to be passed for all possible node pairs in a graph)
- Global efficiency: the mathematical inverse of the path length (i.e., an index for how efficiently information is exchanged in the graph network)
- Community structure: the tendency of nodes to form locally connected clusters representing network organization-forming modules
- Degree: the number of edges each node possesses
- Measures of centrality:
  - Closeness centrality: inverse of the average path length; the distance (i.e., numbers of nodes that have to be passed) from one node to every other node in the network;
  - Betweenness centrality: how often a node is used as a "bridge" (i.e., it signals how much that particular node works as a link between networks)

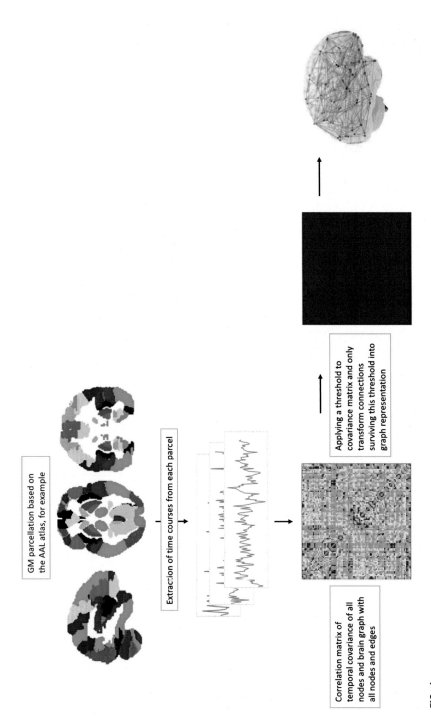

FIG. 1

Step-by-step schematic of the graph theoretical analysis of functional connectivity using the example of gray matter (GM) parcellation based on the Automated Anatomical Labelling atlas; time courses are extracted from each anatomical region, then cross-correlated; thresholding the covariance matrix yields connections that are transformed into graph theoretical metrics.

Connectivity matrices and brain graph from Wang, J., Zuo, X., He, Y., 2010. Graph-based network analysis of resting-state functional MRI. Front. Syst. Neurosci. 4:16. https://doi.org/10.3389/fnsys.2010.00016.

Further metrics can be derived from the combination of the previously mentioned measures. For example, a node with many links (high degree) and with high centrality yields a *hub* [Rubinov and Sporns, 2010]). Those hubs are again strongly interconnected among themselves forming a central rich club (van den Heuvel and Sporns, 2011). It is assumed that the brain's rich club represents a central backbone of FC, and that it plays a central role in integrating information from different functional domains. Anatomically, this rich club encompasses the precuneus, superior frontal, and superior parietal cortices, as well as subcortical brain regions such as the hippocampus, putamen, and thalamus (van den Heuvel and Sporns, 2011). Interestingly, this partially resembles the DMN with additional regions from the TPN, salience, and sensory-motor domains.

Findings in schizophrenia

Overall, patients with schizophrenia tend to show longer path lengths, reduced centrality, and reduced rich club interconnectivity (Bassett et al., 2008; Kambeitz et al., 2016; Lynall et al., 2010; van den Heuvel et al., 2013; van den Heuvel and Fornito, 2014). Anatomically, these alterations are particularly observed between frontal and parietal brain regions. These findings suggest that functional segregation, as reflected by centrality and functional integration (path length and rich club interconnectivity, for example), are impaired in schizophrenia according to graph theoretical approaches. As a proof of disease specificity, these alterations have further been related to symptom severity. A positive correlation was found between the path length and the negative score of the Positive and Negative Symptom Scale (PANSS, [Kay et al., 1987]) by Yu et al. (2011), where they observed a negative correlation between the negative PANSS score and global efficiency. Further, Wang et al. (2012) showed that the total, positive, and negative scores of the PANSS were negatively correlated with connectivity of hubs, specifically in the prefrontal cortex, (para-)limbic, and subcortical brain regions. Another study found that more pronounced hub dysconnectivity precedes the cognitive decline in patients with schizophrenia (Collin et al., 2016). Other work focusing on the use of ICA to derive network nodes for graph analyses have shown differences in modular organization in schizophrenia (Yu et al., 2012) and links between connectivity strength and low frequency power (Yu et al., 2013). A simulation study recently suggested that the use of ICA is more accurate than an ROI approach, as it does not assume that ROIs are identical across voxels or subjects (Yu et al., 2017).

As we will further discuss in the context of ICA-based connectivity approaches, studies have recently started to take dynamic changes of FC into account. With regard to graph theory, Yu et al. (2015) suggested a sliding-window framework to investigate changes of graph metrics instead of summarizing these across the entire time course of the resting-state scan. In this study, they found that patients with schizophrenia exhibited lower variance of dynamic graph metrics and showed reduced connectivity strength, clustering, and global efficiency across all identified connectivity states.

Overall, these findings underline the early notion by Bleuler claiming that schizophrenia can be viewed as a disorder in which information from different functional domains are erroneously integrated, leading to a decay of higher cognitive functions that rely in particular on integration of inputs from different brain regions.

Strengths and weaknesses

A strength of graph theoretical approaches is the relatively straightforward interpretation of network summary measures as well as the comparisons of those between groups. Further, physiological features of brain connectivity strongly suggest a network-like organization, and applying methods stemming from network sciences seems logical.

However, Graph theoretical measures of connectivity rely on brain parcellation techniques. The vast number of different parcellation techniques, such as atlas-based, functional separation or data-driven approaches, highlights one of the limitations of this approach. Not only the number of nodes included in a study but also the anatomical region believed to be represented by that node varies from study to study depending on the parcellation algorithm applied. This renders it difficult to achieve replicability of findings (Hallquist and Hillary, 2018). Further, correlation algorithms applied to calculate the connectivity matrix vary where some studies use bivariate and others partial correlation (Varoquaux and Craddock, 2013) or wavelet decomposition approaches (Chai et al., 2017; Hutchison et al., 2013). Along the same lines, there is also no consistent way as to how researchers translate negative correlations into graph metrics. Even though these issues are a matter of ongoing research, clinical studies often neglect these issues (Rubinov and Sporns, 2010).

ROI-/SEED-BASED FUNCTIONAL CONNECTIVITY

The seed-based approach was the first method used to investigate FC as it is easy to implement and straightforward to interpret, but clear hypotheses about which particular brain regions to investigate are required.

Technical foundations

To perform seed-based connectivity analysis, resting-state fMRI data has to be preprocessed including thorough processes to eliminate subjects' motion (Power et al., 2014). The selection of a seed can be based on anatomy, prior study results, or a functional localizer (Friston et al., 2006; Poldrack, 2007). Once a seed region is selected and defined, time courses from that seed region are extracted from preprocessed resting-state fMRI data. To calculate whole-brain voxel-wise FC maps of cross-correlation with the seed region, the time courses are either used as a regressor in a linear correlation analysis, or modeled with another confounding regressor of no interest in a general linear model (GLM) framework. Finally, after performing statistical tests to identify group differences and accounting for multiple comparisons correction, significant connections between the seed and other brain areas can be identified. Steps involved in performing this analysis are presented in Fig. 2.

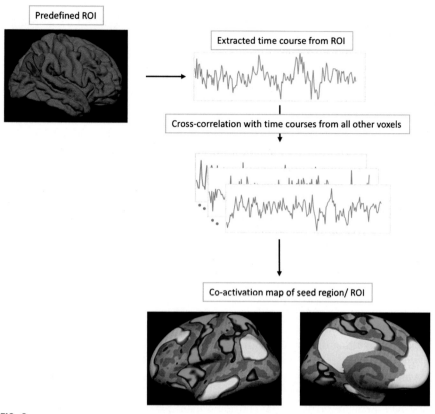

FIG. 2

Step-by-step schematic of region of interest (ROI)-based analysis of functional connectivity; time courses are being extracted from a predefined region followed by cross-correlation with all other voxels in the brain; setting an appropriate threshold yields regions that are functionally connected with this particular ROI.

Findings in schizophrenia

With regard to schizophrenia, considerable emphasis has been put on prefrontal/insular regions, areas associated with the DMN, as well as subcortical structures such as the thalamus and putamen. The rationale for choosing these regions is based on findings in PET and task-based studies. For example, Lawrie et al. (2002) showed in their seed-based analysis using task-based fMRI data of a sentence completion paradigm that connectivity between the left dorsolateral prefrontal cortex and left superior temporal gyrus is reduced in patients with schizophrenia. In another task-based connectivity study, Boksman et al. (2005) found that patients with a first episode of schizophrenia lack interaction between the anterior cingulate cortex and the temporal lobe in a word fluency task. Whitfield-Gabrieli et al. (2009) studied seed-based connectivity in a working memory task in a sample of patients with schizophrenia and their first-degree relatives; in both populations, they found hyperconnectivity between prefrontal DMN and TPN brain areas. Using bigger seeds (ROI) on resting-state fMRI data, Woodward et al. (2012) found prefrontal-thalamic hypoconnectivity vs. motor-somatosensory-thalamic hyperconnectivity in patients with schizophrenia. Aiming at exploring whole-brain connectivity and using whole-brain parcellation (Tzourio-Mazoyer et al., 2002) instead of seeds, Liang et al. (2006) identified widespread dysconnectivity with increased connectivity in the cerebellum and decreased connectivity among the insular cortex, the temporal and prefrontal lobe, and the striatum. Based on the same parcellation algorithm, Klingner et al. (2014) specified previously reported subcortical dysconnectivity. In their study, patients with schizophrenia exhibited overall increased thalamic connectivity but in particular with ventrolateral prefrontal regions, motor cortices, and temporal and occipital brain regions, which partially contradicts findings from Woodward et al. (2012).

In general, findings from seed-/ROI-based connectivity analyses summarize a large variety of functional dysconnectivity in schizophrenia, which varies considerably based on the choice of specific seeds/ROIs.

Strengths and weaknesses

Relative to other FC analysis techniques, the main strength of the seed-based approach is its straightforwardness, both in terms of implementation and result interpretation. Another advantage of this approach lies in its test-retest reliability, with evidence suggesting that it can identify patterns of connectivity among resting-state networks with moderate to high reliability (Shehzad et al., 2009).

One of the major limitations of the seed-/ROI-based approach is the requirement of a priori selection of a seed voxel or atlas region, typically defined by previously published literature or functional activation maps extracted from previous studies. Also, seed time courses are typically analyzed within a univariate statistical model, as data from each voxel is regressed against the model of choice. This step is separately performed for every voxel, not capitalizing on the vast information available within the temporal dependencies among multiple data points. Seed-based studies have an enormous value when testing very specific hypotheses but fail to provide an overarching synopsis on whole-brain connectivity and patterns of dysconnectivity in schizophrenia in particular.

INDEPENDENT COMPONENT ANALYSIS

Technical foundations—Static functional network connectivity

ICA is a data-driven technique that uses higher order statistics to separate signals from multiple hidden sources and is viewed as a special case of blind source separation. The complexity of neuroimaging data, with multiple thousands of voxels and associated time course, is well-suited to ICA (Beckmann and Smith, 2005; Calhoun et al., 2001a,b; McKeown et al., 1998). Independence with respect to resting-state fMRI data commonly refers to the spatial dimension, and the goal is to separate brain regions with coupled time courses (i.e., independent components [Calhoun et al.,

2001a,b; McKeown et al., 1998]). Compared with the approaches discussed earlier in this chapter, the preprocessing of resting-state fMRI data for ICA is less complex. This is due to the fact that, by separating data into independent components, ICA, also identifies sources of noise. Motion removal and bandpass filtering is typically applied after ICA to preserve more variance in the data to facilitate the ICA algorithm. With regard to resting-state fMRI connectivity analysis, ICA is often performed on concatenated data from all subjects in the sample (after data reduction steps involving principle component analysis), which is called group ICA (GICA [Calhoun et al., 2001a,b; Calhoun and Adali, 2012; Correa et al., 2007]). The output of such a spatial GICA comprises the group's mean spatial maps and according time courses for all independent components. Spatial maps and time courses are used to back-reconstruct a single subject's individual spatial maps and time courses. Multiple studies have shown the reliability of the GICA approach and that it can preserve individual differences (Allen et al., 2012b; Biswal et al., 2010; Calhoun and Adali, 2012; Du and Fan, 2013; Erhardt et al., 2011). The model order of the group ICA approach reflects in how many independent components or sources the data is separated into and has a large influence on the ICA decomposition when under- or overestimated (Allen et al., 2012b). Many studies these days use a higher model order (i.e., 100 independent components). Independent components identified as conveying nonartifactual information—ICNs—are further used to calculate the covariance of all ICNs among each other yielding a static functional network connectivity (FNC) matrix. FNC in contrast to FC is defined as the temporal covariation across ICNs (Jafri et al., 2008). The most commonly used metric to calculate FNC is the Pearson correlation coefficient. Investigations of group differences can be performed on correlation values between ICNs, activation differences in spatial maps, or frequency fluctuations in time courses (Zou et al., 2008). Steps involved in group ICA, and static and dynamic FNC are summarized in Fig. 3.

STATIC FUNCTIONAL NETWORK CONNECTIVITY
Findings in schizophrenia
One of the initial studies that implemented ICA to analyze task-based fMRI data extracted specifically default mode components and examined their spatial and temporal features, as well as their FNC (Garrity et al., 2007). Garrity and colleagues found increased DMN activity during a working memory task in patients with schizophrenia. Further, patients exhibited increased power in higher frequencies, indicative of less temporal synchronicity. Another schizophrenia-based static FNC study used resting-state fMRI to evaluate the connectivity between DMN components (extracted using ICA) and each voxel in the whole brain, and assessed the groups for differences in coupling with the default mode in terms of DMN correlation coefficients (Camchong et al., 2011). Again, this study found hyperconnectivity in prefrontal brain areas. However, these studies were mainly focused on DMN connectivity. Even though the DMN appears to play a pivotal role in the synchronized functioning of the brain, it is also important to investigate the brain regions the DMN is interacting with,

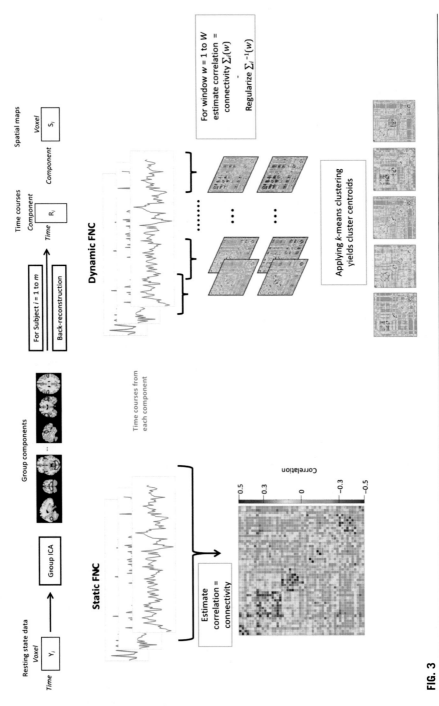

FIG. 3

Step-by-step schematic of group independent component analysis of functional network connectivity (FNC); first, group independent component analysis decomposes the data, yielding spatial maps with associated time courses; for static FNC, a covariance matrix is created based on nonartifactual components; for dynamic FNC, a sliding-temporal window splits time courses in smaller chunks clustered around recurring whole-brain connectivity patterns.

suggesting that a whole-brain approach is favorable. With respect to schizophrenia, the whole-brain static FNC approach was first implemented in 2008 by Jafri and colleagues (Jafri et al., 2008). In their study, Jafri et al. used a low model order of 30 independent components of which seven were chosen as ICNs of interest. These seven components were assigned to the following functional domains: DMN, visual domain, fronto-temporo-parietal domain, fronto-temporo-subcortical domain, parietal domain, frontal domain, and temporal domain. Then, connectivity between components was estimated. Results from this study indicate that patients with schizophrenia showed increased connectivity between DMN components and frontal and visual domains compared with healthy controls. Interestingly, a recent classification study (Arbabshirani et al., 2013) used features, specifically correlation coefficients of ICN pairs, from static FNC to discriminate patients with schizophrenia from healthy controls at an individual level, where they identified nine independent components as resting-state networks from a model order of 30 components. The nine resting-state networks were grouped into attentional network, visual network, DMN, motor network, and auditory network, and static FNC metrics were computed as the pairwise cross-correlations between the components' time courses. Using these static FNC metrics for healthy controls and patients with schizophrenia, the authors proposed a classification framework. Both linear and nonlinear classifiers were trained and evaluated based on their classification accuracy performance. Significant performance in terms of classification accuracy was achieved using the nonlinear classifier such as k-nearest neighbor method. Shen et al. (2010) used an atlas-based method to extract mean time-courses of 116 brain regions in the resting state for both healthy controls and schizophrenia subjects. The correlation between these time courses made the feature vector for each subject. By applying feature selection and dimensionality reduction methods, they reduced the dimensionality down to three where they classified patients from controls with a high accuracy. Calhoun et al. (2008) used the temporal lobe and DMNs as features using a leave-one-out cross-validation framework, and classified schizophrenia and bipolar patients at the individual level. Therefore, future classification studies based on resting-state fMRI connectivity features might offer powerful tools that can potentially discriminate schizophrenia from other mental disorders sharing overlapping symptoms.

Dynamic functional network connectivity

As previously mentioned, the assumption that FC is static over a longer time period does not appear intuitive because neuronal processes are constantly present, rendering changes in connectivity likely, and we will now further elaborate this consideration. As the name implies, static FNC approaches capture only the average or steady-state connectivity of the brain and neglect the fact that individual subjects are likely to engage in slightly different mental activities at different instances in time (Arieli et al., 1996; Makeig et al., 2004). The assumption of stationarity was further challenged in a recent work focused on time-varying multivariate connectivity patterns (Sakoğlu et al., 2010). Several following studies suggest that FC changes over a short period of time of seconds to minutes (Allen et al., 2018, 2012a; Matsui et al., 2018). Moreover, studies have demonstrated the utilization of the information contained within the temporal features of spontaneous fluctuation of BOLD signals (Hutchison et al., 2013) providing results that could not have been detected with static FNC analyses. Thus, a more advanced ICA-based approach to analyze FC of the brain is dynamic FNC, which is designed to capture dynamically changing connectivity between synchronously activated brain components (Allen et al., 2012a; Calhoun et al., 2014). The dynamic FNC approach efficiently unpacks the average connectivity captured by static FNC into different meaningful connectivity states. By learning about various mental states experienced during rsfMRI scans, time spent in those states, and transitions between states, we begin to further understand modulations and features of FNC. Therefore, looking beyond the static FNC and assessing dynamic changes in connectivity seems favorable.

Moreover, in static FNC analysis, the focus is most often solely on component-level patterns of connectivity across a massive matrix of information. In contrast, in time-varying dynamic FNC analysis, additional numerical summary metrics can be calculated, such as the state transition matrix reflecting likelihoods of transitioning from one state to any other state, mean dwell time (MDT, i.e., the time a subject spends in one state before transitioning to another one), fraction of time (FT, i.e., cumulative time spend in one state across the entire resting state scan), and the occupancy rate. These metrics allow for a quantitative interpretation of a given state as a whole beyond the mere interpretation of component-level connectivity patterns (Allen et al., 2012a). These quantitative summary measures aid in simplifying interpretations of complex network information, which historically are often subjectively evaluated.

Technical foundations—Dynamic FNC

The most widely used approach to estimate connectivity dynamics is the sliding-window approach. For this type of analysis, time courses of independent components derived from GICA are parceled into smaller temporal "chunks" by letting a temporal window slide across the entire time course. FNC within each window is—just like the static FNC approach—quantified by the cross-correlation among ICNs. In the dynamic FNC approach, a time window of fixed length (preferable with tapered edges) is selected, and data from time points within that window are used to calculate the FNC of that particular window. The window is then shifted in time by a fixed number of time points (ranging from a single time point to the length of a window) that defines the amount of overlap between consecutive windows. The most commonly used metric to calculate FNC is the Pearson correlation coefficient. The time-varying behavior of the chosen metric over the scan period can then be quantified and summarized. Any metric used to estimate static FNC can be applied to the sliding-window analysis, given that there is a sufficient number of time points for robust calculation. The resulting windowed FNC matrices are further clustered (e.g., by k-means clustering [Steinley, 2006]) to identify recurring whole-brain connectivity patterns across the resting-state scan.

The dynamic analysis of resting-state FC is still relatively young within the neuroimaging research community that involves lively debates encouraging further research on alternatives to the sliding-window approach, such as wavelet-based decomposition (Chai et al., 2017; Chang and Glover, 2010; Hutchison et al., 2013; Yaesoubi et al., 2015), dynamic conditional correlation (Lindquist et al., 2014), windowless approaches (Yaesoubi et al., 2018), improvement of statistical metrics applied, and development of appropriate null models (Hindriks et al., 2016; Miller et al., 2017; Shakil et al., 2016; Smith et al., 2013; Zalesky and Breakspear, 2015).

Findings in schizophrenia

With respect to schizophrenia, published studies are just emerging. In Damaraju et al. (2014), the sliding-window-based dynamic FNC approach was implemented, and group differences in patients with schizophrenia and healthy subjects were observed. Results from this study showed hyperconnectivity between thalamus and sensorimotor networks, as well as hypoconnectivity within the sensorimotor networks. Moreover, group differences were more prominent in states that showed stronger negative cortical-subcortical connectivity, an important addition to what static FNC analyses had found before. In a combined ICA-graph theoretical approach, Du et al. (2016) particularly explored dynamic FNC in patients with schizophrenia within the DMN domain, showing that, compared with healthy individuals, patients spent less time in tightly connected states and exhibited reduced average node strength, clustering coefficient, and both global and local efficiency. These results reflected impairment in DMN subsystems, specifically in the posterior cingulate cortex and anterior medial prefrontal cortex, as well as reduced connectivity between dorsal medial prefrontal cortex and medial temporal lobe.

Another important possible application of the dynamic FNC approach lies in clinical diagnostics. A recent study (Rashid et al., 2014) has shown evidence that changes in connectivity dynamics can explain underlying similarities and differences in schizophrenia and bipolar disorder patients, providing a scope for biological markers related to the illness. Rashid et al. (2016) showed that incorporating features from window-based dynamic FNC analyses for subject-level classification provides more sensitive and disease-specific markers than static FNC features, which can reveal connectivity differences between two patient groups sharing overlapping symptoms, such as patients with schizophrenia and patients with bipolar disorder. Also, disease-specific markers as measured by differences in dynamic FNC between healthy controls and patient groups were mapped onto specific dynamic states, capturing both hypo- and hyperconnectivity for the patient groups compared with baseline connectivity in healthy controls. Anatomically, group differences encompassed connectivity among DMN, temporal, and frontal components, supporting the contribution of aberrant DMN and temporal lobe components' connectivity in schizophrenia.

To characterize the dynamic FNC features as temporal summary measures and mitigate the dimensionality issue associated with the data per se, Miller et al. (2016) proposed a so-called meta-state approach. Within this framework, windowed connectivity is formulated in terms of distance vectors to each state's main connectivity pattern acquired by organizing data with approaches such as k-means clustering, spatial ICA, or principle component analysis; this approach is visualized in Fig. 4.

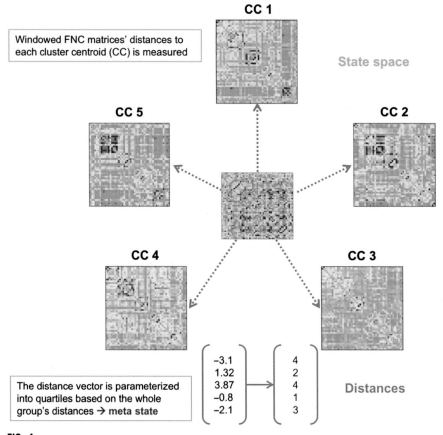

FIG. 4

Schematic of the metastates approach; the same steps are involved as in dynamic FNC analysis up to the clustering; after yielding cluster centroids for each state, windowed FNC matrices are not sorted to one and only one cluster centroid, but rather the distance to each one of the cluster centroids is measured yielding a vector that is then parametrized; the parametrized vector is called metastate and further measure of temporal dynamism can be derived from all metastates.

This approach allows for connectivity states to overlap in time and bypasses that each windowed FNC matrix is forced into one and only one state membership. The parameterized distance vector of each windowed FNC matrix representing the overlapping states at a specific time is called a meta-state. From this vector, summary measures, such as number of unique meta-state and meta-state changes, can be derived. Results from Miller et al. showed that patients with schizophrenia displayed less overall temporal dynamism as they exhibited fewer changes in overall connectivity patterns compared with healthy subjects.

Such observations suggest that dynamic FNC analyses reveal more detailed information about (dys)connectivity in health and disease, and that these approaches also have the ability to discriminate between patients with different disorders, an application that might become useful to diagnose complex psychiatric illnesses or disorders with similar symptoms that can only be clinically diagnosed based on the course of the disorder itself.

Strengths and weaknesses

One of the challenges, in particular with regard to dynamic FNC analysis, is that to robustly estimate the covariance matrix, enough time points should be obtained in each temporal window. Especially when a short window length is chosen, noise might be mistaken as meaningful dynamic connectivity changes. On the other hand, sensitivity to pick up dynamic changes decreases with very long window size (Lindquist et al., 2014; Sakoğlu et al., 2010). Moreover, acquiring longer resting-state scans is preferable to study the complexity of dynamic connectivity states by assessing more temporal windows with a sufficient window length, although the identification of an optimal scan duration is still an open issue. Another major concern is the choice of window size. All studies reviewed applied a predefined window length. A rarely used alternative is an approach that estimates the optimal window length based on the data (Ombao and Bellegem, 2008).

An advantage of ICA-based approaches is that resting-state networks identified by ICA tend to be less prone to noise-driven artifactual effects (including fluctuations in the mean global signal) than those from seed-based approaches due to the ability of the method to account for the existence of such structured noise effects within additional ICA components (Biswal et al., 2010; Calhoun and Adali, 2012). Replicability of dynamic FNC states and the basic GICA approach has been shown previously in a large sample of 7500 subjects (Abrol et al., 2017).

In a comparatively unrestricted way, the ICA-based dynamic FNC approach has been used to generate a "more complete" picture of the functional association and segregation among various brain networks that might underlie the spontaneous neuronal fluctuation of the human brain.

Functional connectivity in the clinical high-risk population

As previously discussed, schizophrenia usually does not start abruptly but is rather preceded by a prodromal period characterized by attenuated psychotic symptoms not yet fulfilling diagnostic criteria for schizophrenia. This state in particular offers unique opportunities to investigate neural underpinnings and pathophysiology in schizophrenia. Moreover, we have validated tools to diagnose this clinical high-risk (CHR) state (McGlashan et al., 2010), and more studies seek to investigate alterations prevalent in CHR individuals. For the sake of completeness, we will briefly summarize findings in this specific group across the different connectivity approaches previously discussed.

With regard to graph theoretical FC approaches, recent work suggests that CHR individuals exhibit reduced global efficiency compared with healthy control individuals. However, this decrease is not as pronounced as in schizophrenia patients (Choi et al., 2017). Applying a novel group ICA approach, Du et al. (2017) observed something very similar where dysconnectivity in the CHR sample was not as pronounced as in patients with schizophrenia. Nevertheless, CHR individuals also exhibited unique patterns of alterations. Moreover, results from seed-based studies and two major brain foci of these (i.e., thalamic and DMN dysconnectivity) further support the notion of present but less pronounced alterations in CHR individuals compared with healthy controls (Anticevic et al., 2015; Colibazzi et al., 2017; Wotruba et al., 2013). Most of these studies do not report longitudinal data, and it is not possible to differentiate between two explanations of the observed effects: either CHR individuals indeed show alterations of FC that are not as pronounced as in patients with schizophrenia, or just the subgroup of CHR individuals who will eventually transition to full-blown psychosis exhibits alterations to the same extent as patients with schizophrenia, but this effect is diluted by the majority of CHR individuals who will not transition to full-blown psychosis. Longitudinal studies are needed to test these two alternatives and eventually identify brain-derived biomarkers that would help to predict future development of psychosis in those individuals at risk. However, relatively low transition rates to full-blown psychosis within this high-risk sample (circa 30% of individuals in this group over a 2-year period) raise doubt on the overall validity of the CHR diagnostic construct (van Os and Guloksuz, 2017).

SUMMARY

Functional dysconnectivity in schizophrenia can be identified with many different approaches, each one of them offering unique advantages but also coming with specific shortcomings. For seed-/ROI-based studies, findings are highly dependent on the chosen regions and the preprocessing steps (e.g., filtering and covariation with physiological variables). Moreover, despite correction for multiple comparisons, the amount of performed tests in such an analysis is vast. Nevertheless, seed-based approaches are very useful when testing specific hypotheses or when combining task-based and resting-state fMRI data to scrutinize task-associated connectivity.

In contrast to seed-based studies, ICA is a purely data-driven approach that does not require a priori selection of which brain region to investigate but rather identifies brain regions based on the data themselves. However, the limiting factors of this approach are the uncertainty with regard to the model order (i.e., the number of independent components estimated) and the fact that every time an ICA is performed, different independent components are extracted resulting in diminished comparability across studies. Even though these limitations are present, replicable reoccurring brain regions and connectivity patterns across various studies have been reported, suggesting ICA as a powerful analysis tool (Abrol et al., 2017; Calhoun and Adali, 2012).

Graph theory approaches provide global summary metrics. However, establishing the direction among connections can be challenging, as association between brain regions does not necessarily correspond to direct connection. Most commonly, nodes are based on atlas-derived parcellations, which secures comparability on the one hand but, on the other hand, constrains the analysis similar to ROI- and seed-based approaches. The combination of ICA with graph theory can mitigate this issue.

Despite methodological differences between these approaches, commonalities with respect to dysconnectivity in schizophrenia are present.

(a) Anatomical overlap: Even though dysconnectivity can be identified in many different brain areas, some regions overlap across different approaches, even in data-driven approaches without spatial constraints. In particular, anterior and posterior cores of the DMN, thalamus, and lateral prefrontal brain areas seem to be hot spots of dysconnectivity in schizophrenia. Graph theoretical approaches have identified reduced rich club connectivity in patients with schizophrenia, indicating that the central structure orchestrating communication between functional networks, namely the DMN, is disrupted in those patients (van den Heuvel et al., 2013). Importantly, even though hot spots of dysconnectivity can be identified, these areas exhibit functional connections to multiple brain areas highlighting the importance of whole-brain connectivity approaches.

(b) Local hypo- and hyperconnectivity: Compared with healthy subjects, schizophrenia patients exhibit local hypo- and hyperconnectivity. Hypoconnectivity (i.e., the correlation of time courses between regions is reduced in patients with schizophrenia compared with healthy control individuals) is present within sensorimotor brain areas as well as in cerebellar-cortical and -subcortical connections (Anticevic et al., 2014; Damaraju et al., 2014; Kim et al., 2014; Shinn et al., 2015). Hyperconnectivity, on the other hand, is found between DMN and prefrontal TPN brain areas, and between the thalamus and sensorimotor brain areas (Anticevic et al., 2014; Whitfield-Gabrieli et al., 2009; Whitfield-Gabrieli and Ford, 2012).

(c) Less dynamic connectivity: Patients with schizophrenia exhibit more rigid patterns of whole-brain connectivity as indicated by reduced variability of connectivity across graph theoretical and group ICA approaches (Miller et al., 2016; Yu et al., 2015). Patients with schizophrenia manifest less dynamism by occupying only a limited amount of available connectivity states, and they also exhibit reduced variability in graph metrics.

In this chapter, we reviewed results from highly focused studies in very particular branches of brain connectivity research. Even though interesting and important findings can be distilled from these studies, the summary yields an unsatisfying overall picture. For future research, it therefore appears pivotal to advance and exploit multi-modal fusion approaches that truly capitalize on the rich information contained in the different types of data that we typically have at hand such as structural and functional imaging data, genetic data, and behavioral measures (Calhoun and Sui, 2016; Sui et al., 2013, 2015). These promising multivariate approaches identify interrelated patterns of disease-related effects across modalities that otherwise would have been hidden in the multidimensionality of the data and that might have been too small to be detected (Calhoun and Sui, 2016).

FUTURE DIRECTIONS

One of the major goals in psychiatric research is to identify underlying neural patho-physiological correlates of symptoms to yield a holistic understanding of specific diseases, thereby enabling development of new treatment approaches, and establishment of personalized treatment planning.

With regard to schizophrenia, the hope is to be able to provide validated fMRI-derived metrics to:

(a) Classify disorders presenting with similar clinical symptoms such as schizophrenia and bipolar disorder with psychotic symptoms early on. Both illnesses are currently diagnosed using a combination of cross-sectional clinical symptoms, the longitudinal course of symptoms, and outcome measures. Especially shortly after the onset of symptoms, it is difficult and sometimes even impossible to clearly diagnose a patient because the symptom trajectory cannot be foreseen. With neuroimaging tools and techniques discussed in this chapter, it might be possible to establish a set of reliable disease-specific biomarkers, which may aid in developing an accurate and reliable diagnostic platform that can help to guide treatment. Discerning between these two illnesses is important to prescribe appropriate medication. In the case of bipolar disorder, medication includes mood stabilizers first and foremost, whereas first-line pharmacological treatment of schizophrenia encompasses antipsychotics.

(b) Predicting future symptom development would also help in the treatment of CHR individuals who present with symptoms not (yet) fulfilling diagnostic criteria for psychosis and their symptom progress is even harder to predict. Applying current diagnostic tools established to diagnose the CHR state for psychosis yields transition rates to full-blown psychosis of roughly 30% over a 2-year period (Cannon, 2015; Cannon et al., 2016). This indicates that there is a need to bolster up clinical measures by biologically informed measures. Being able to better predict full-blown psychosis onset would help to focus clinical efforts on those individuals most likely to develop schizophrenia, which has been shown to delay transition and improve outcome measures (Bechdolf et al., 2012; McGorry et al., 2009; Ruhrmann et al., 2014).

(c) On the other hand, we suggest that these possible streams of research should consider the RDoC framework (Insel et al., 2010) that we mentioned at the beginning of this chapter. Associations between specific symptoms and dysconnectivity patterns should be scrutinized before proceeding to the "interaction" of symptoms yielding specific disease categories. However, pursuing a dimensional approach when studying psychopathology is less intuitive to clinicians who use categorical classifications on a daily basis to make diagnostic decisions. Moreover, a dimensional approach is still lacking a generally accepted construct (Kotov et al., 2018). A solution to this dilemma might be a hybrid of categorical and dimensional approaches.

(d) Moving beyond merely mapping group differences, multimodal approaches combining imaging, genetic, and behavioral data should be used more frequently, capitalizing on multivariate statistical approaches that can identify patterns across modalities. This type of approach can facilitate identifying concrete causes of psychopathology by revealing gene-brain-behavior interactions, for example.

REFERENCES

Abrol, A., Damaraju, E., Miller, R.L., Stephen, J.M., Claus, E.D., Mayer, A.R., Calhoun, V.D., 2017. Replicability of time-varying connectivity patterns in large resting state fMRI samples. Neuroimage 163, 160–176.

Adams, R.A., Stephan, K.E., Brown, H.R., Frith, C.D., Friston, K.J., 2013. The computational anatomy of psychosis. Front. Psych. 4, 47.

Allen, E.A., Damaraju, E., Plis, S.M., Erhardt, E.B., Eichele, T., Calhoun, V.D., 2012a. Tracking whole-brain connectivity dynamics in the resting state. Cereb. Cortex 24, 663–676.

Allen, E.A., Erhardt, E.B., Wei, Y., Eichele, T., Calhoun, V.D., 2012b. Capturing inter-subject variability with group independent component analysis of fMRI data: a simulation study. Neuroimage 59, 4141–4159.

Allen, E.A., Damaraju, E., Eichele, T., Wu, L., Calhoun, V.D., 2018. EEG signatures of dynamic functional network connectivity states. Brain Topogr. 31 (1), 101–116.

Anticevic, A., Cole, M.W., Repovs, G., Murray, J.D., Brumbaugh, M.S., Winkler, A.M., Savic, A., Krystal, J.H., Pearlson, G.D., Glahn, D.C., 2014. Characterizing Thalamo-cortical disturbances in schizophrenia and bipolar illness. Cereb. Cortex 24, 3116–3130.

Anticevic, A., Haut, K., Murray, J.D., Repovs, G., Yang, G.J., Diehl, C., McEwen, S.C., Bearden, C.E., Addington, J., Goodyear, B., Cadenhead, K.S., Mirzakhanian, H., Cornblatt, B.A., Olvet, D., Mathalon, D.H., McGlashan, T.H., Perkins, D.O., Belger, A., Seidman, L.J., Tsuang, M.T., van Erp, T.G.M., Walker, E.F., Hamann, S., Woods, S.W., Qiu, M., Cannon, T.D., 2015. Association of thalamic dysconnectivity and conversion to psychosis in youth and young adults at elevated clinical risk. JAMA Psychiat. 72, 882.

Arbabshirani, M.R., Kiehl, K.A., Pearlson, G.D., Calhoun, V.D., 2013. Classification of schizophrenia patients based on resting-state functional network connectivity. Front. Neurosci. 7, .

Arieli, A., Sterkin, A., Grinvald, A., Aertsen, A., 1996. Dynamics of ongoing activity: explanation of the large variability in evoked cortical responses. Sci. Wash. 273, 1868.

Balu, D.T., 2016. The NMDA receptor and schizophrenia: from pathophysiology to treatment. In: Schwarcz, R. (Ed.), Advances in Pharmacology, Neuropsychopharmacology: A Tribute to Joseph T. Coyle. Academic Press, San Diego, CA, pp. 351–382. (Chapter 12).

Bassett, D.S., Bullmore, E., Verchinski, B.A., Mattay, V.S., Weinberger, D.R., Meyer-Lindenberg, A., 2008. Hierarchical Organization of Human Cortical Networks in health and schizophrenia. J. Neurosci. 28, 9239–9248.

Bechdolf, A., Wagner, M., Ruhrmann, S., Harrigan, S., Putzfeld, V., Pukrop, R., Brockhaus-Dumke, A., Berning, J., Janssen, B., Decker, P., Bottlender, R., Maurer, K., Moller, H.-J., Gaebel, W., Hafner, H., Maier, W., Klosterkotter, J., 2012. Preventing progression to first-episode psychosis in early initial prodromal states. Br. J. Psychiatry 200, 22–29.

Beckmann, C.F., Smith, S.M., 2005. Tensorial extensions of independent component analysis for multisubject FMRI analysis. Neuroimage 25, 294–311.

Biswal, B., Zerrin Yetkin, F., Haughton, V.M., Hyde, J.S., 1995. Functional connectivity in the motor cortex of resting human brain using echo-planar MRI. Magn. Reson. Med. 34, 537–541.

Biswal, B.B., Mennes, M., Zuo, X.-N., Gohel, S., Kelly, C., Smith, S.M., Beckmann, C.F., Adelstein, J.S., Buckner, R.L., Colcombe, S., Dogonowski, A.-M., Ernst, M., Fair, D., Hampson, M., Hoptman, M.J., Hyde, J.S., Kiviniemi, V.J., Kötter, R., Li, S.-J., Lin, C.-P., Lowe, M.J., Mackay, C., Madden, D.J., Madsen, K.H., Margulies, D.S., Mayberg, H.S., McMahon, K., Monk, C.S., Mostofsky, S.H., Nagel, B.J., Pekar, J.J., Peltier, S.J., Petersen, S.E., Riedl, V., Rombouts, S.A.R.B., Rypma, B., Schlaggar, B.L., Schmidt, S., Seidler, R.D., Siegle, G.J., Sorg, C., Teng, G.-J., Veijola, J., Villringer, A., Walter, M., Wang, L., Weng, X.-C., Whitfield-Gabrieli, S., Williamson, P., Windischberger, C., Zang, Y.-F., Zhang, H.-Y., Castellanos, F.X., Milham, M.P., 2010. Toward discovery science of human brain function. Proc. Natl. Acad. Sci. 107, 4734–4739.

Bleuler, E.P., 1908. Die Prognose der Dementia praecox (Schizophreniegruppe). Allg. Z. Fuer Psychiatr. Psych. Med. 65, 436–464.

Bleuler, E.P., 1911. Dementia praecox oder Gruppe der Schizophrenien. . Handbuch der Psychiatrie.

Boksman, K., Théberge, J., Williamson, P., Drost, D.J., Malla, A., Densmore, M., Takhar, J., Pavlosky, W., Menon, R.S., Neufeld, R.W.J., 2005. A 4.0-T fMRI study of brain connectivity during word fluency in first-episode schizophrenia. Schizophr. Res. 75, 247–263.

Calhoun, V.D., Adali, T., 2012. Multisubject independent component analysis of fMRI: a decade of intrinsic networks, default mode, and neurodiagnostic discovery. IEEE Rev. Biomed. Eng. 5, 60–73.

Calhoun, V.D., Sui, J., 2016. Multimodal fusion of brain imaging data: a key to finding the missing link(s) in complex mental illness. Biol. Psychiatry Cogn. Neurosci. 1, 230–244. Neuroimaging, Brain Connectivity in Psychopathology.

Calhoun, V.D., Adali, T., Pearlson, G.D., Pekar, J.J., 2001a. A method for making group inferences from functional MRI data using independent component analysis. Hum. Brain Mapp. 14, 140–151.

Calhoun, V.D., Adali, T., Pearlson, G.D., Pekar, J.J., 2001b. Spatial and temporal independent component analysis of functional MRI data containing a pair of task-related waveforms. Hum. Brain Mapp. 13, 43–53.

Calhoun, V.D., Maciejewski, P.K., Pearlson, G.D., Kiehl, K.A., 2008. Temporal Lobe and 'default' hemodynamic brain modes discriminate between schizophrenia and bipolar disorder. Hum. Brain Mapp. 29, 1265–1275.

Calhoun, V.D., Miller, R.L., Pearlson, G., Adalı, T., 2014. The chronnectome: time-varying connectivity networks as the next frontier in fMRI data discovery. Neuron 84, 262–274.

Camchong, J., MacDonald, A.W., Bell, C., Mueller, B.A., Lim, K.O., 2011. Altered functional and anatomical connectivity in schizophrenia. Schizophr. Bull. 37, 640–650.

Cannon, T.D., 2015. How schizophrenia develops: cognitive and brain mechanisms underlying onset of psychosis. Trends Cogn. Sci. 19, 744–756.

Cannon, T.D., Rosso, I.M., Hollister, J.M., Bearden, C.E., Sanchez, L.E., Hadley, T., 2000. A prospective cohort study of genetic and perinatal influences in the etiology of schizophrenia. Schizophr. Bull. 26, 351–366.

Cannon, T.D., Yu, C., Addington, J., Bearden, C.E., Cadenhead, K.S., Cornblatt, B.A., Heinssen, R., Jeffries, C.D., Mathalon, D.H., McGlashan, T.H., Perkins, D.O.,

Seidman, L.J., Tsuang, M.T., Walker, E.F., Woods, S.W., Kattan, M.W., 2016. An individualized risk calculator for research in prodromal psychosis. Am. J. Psychiatry 173, 980–988.

Canu, E., Agosta, F., Filippi, M., 2015. A selective review of structural connectivity abnormalities of schizophrenic patients at different stages of the disease. Schizophr. Res. 161, 19–28.

Carpenter, W.T., Schiffman, J., 2015. Diagnostic concepts in the context of clinical high risk/attenuated psychosis syndrome. Schizophr. Bull. 41, 1001–1002.

Chai, L.R., Khambhati, A.N., Ciric, R., Moore, T.M., Gur, R.C., Gur, R.E., Satterthwaite, T.D., Bassett, D.S., 2017. Evolution of brain network dynamics in neurodevelopment. Netw. Neurosci. 1, 14–30.

Chang, C., Glover, G.H., 2010. Time-frequency dynamics of resting-state brain connectivity measured with fMRI. Neuroimage 50, 81–98.

Choi, S.-H., Kyeong, S., Cho, K.I.K., Yun, J.-Y., Lee, T.Y., Park, H.Y., Kim, S.N., Kwon, J.S., 2017. Brain network characteristics separating individuals at clinical high risk for psychosis into normality or psychosis. Schizophr. Res. 190, 107–114.

Cohen, J.R., D'Esposito, M., 2016. The segregation and integration of distinct brain networks and their relationship to cognition. J. Neurosci. 36, 12083–12094.

Colibazzi, T., Yang, Z., Horga, G., Yan, C.-G., Corcoran, C.M., Klahr, K., Brucato, G., Girgis, R.R., Abi-Dargham, A., Milham, M.P., Peterson, B.S., 2017. Aberrant temporal connectivity in persons at clinical high risk for psychosis. Biol. Psychiatry Cogn. Neurosci. Neuroimag. 2, 696–705.

Collin, G., de Nijs, J., Hulshoff Pol, H.E., Cahn, W., van den Heuvel, M.P., 2016. Connectome organization is related to longitudinal changes in general functioning, symptoms and IQ in chronic schizophrenia. Schizophr. Res. 173, 166–173. Progressive Brain Tissue Loss in Schizophrenia.

Correa, N., Adalı, T., Calhoun, V.D., 2007. Performance of blind source separation algorithms for fMRI analysis using a group ICA method. Magn. Reson. Imaging 25, 684–694.

Coyle, J.T., 2012. NMDA receptor and schizophrenia: a brief history. Schizophr. Bull. 38, 920–926.

Damaraju, E., Allen, E.A., Belger, A., Ford, J.M., McEwen, S., Mathalon, D.H., Mueller, B.A., Pearlson, G.D., Potkin, S.G., Preda, A., Turner, J.A., Vaidya, J.G., van Erp, T.G., Calhoun, V.D., 2014. Dynamic functional connectivity analysis reveals transient states of dysconnectivity in schizophrenia. NeuroImage Clin. 5, 298–308.

de Leeuw, M., Kahn, R.S., Vink, M., 2015. Fronto-striatal dysfunction during reward processing in unaffected siblings of schizophrenia patients. Schizophr. Bull. 41, 94–103.

de Leeuw, M., Bohlken, M.M., Mandl, R.C., Hillegers, M.H., Kahn, R.S., Vink, M., 2017. Changes in white matter organization in adolescent offspring of schizophrenia patients. Neuropsychopharmacology 42, 495–501.

Deco, G., Tononi, G., Boly, M., Kringelbach, M.L., 2015. Rethinking segregation and integration: contributions of whole-brain modelling. Nat. Rev. Neurosci. 16, 430–439.

Di Biase, M.A., Zalesky, A., O'Keefe, G., Laskaris, L., Baune, B.T., Weickert, C.S., Olver, J., McGorry, P.D., Amminger, G.P., Nelson, B., Scott, A.M., Hickie, I., Banati, R., Turkheimer, F., Yaqub, M., Everall, I.P., Pantelis, C., Cropley, V., 2017. PET imaging of putative microglial activation in individuals at ultra-high risk for psychosis, recently diagnosed and chronically ill with schizophrenia. Transl. Psychiatry 7, e1225.

Du, Y., Fan, Y., 2013. Group information guided ICA for fMRI data analysis. Neuroimage 69, 157–197.

Du, Y., Pearlson, G.D., Yu, Q., He, H., Lin, D., Sui, J., Wu, L., Calhoun, V.D., 2016. Interaction among subsystems within default mode network diminished in schizophrenia patients: a dynamic connectivity approach. Schizophr. Res. 170, 55–65.

Du, Y., Fryer, S.L., Fu, Z., Lin, D., Sui, J., Chen, J., Damaraju, E., Mennigen, E., Stuart, B., Mathalon, D.H., Calhoun, V.D., 2017. Dynamic functional connectivity impairments in early schizophrenia and clinical high-risk for psychosis. Neuroimage . (in press).

Ecker, D., Unrath, A., Kassubek, J., Sabolek, M., 2009. Dopamine agonists and their risk to induce psychotic episodes in Parkinson's disease: a case-control study. BMC Neurol. 9, 23.

Erhardt, E.B., Rachakonda, S., Bedrick, E.J., Allen, E.A., Adali, T., Calhoun, V.D., 2011. Comparison of multi-subject ICA methods for analysis of fMRI data. Hum. Brain Mapp. 32, 2075–2095.

Filippi, M., Canu, E., Gasparotti, R., Agosta, F., Valsecchi, P., Lodoli, G., Galluzzo, A., Comi, G., Sacchetti, E., 2014. Patterns of brain structural changes in first-contact, antipsychotic drug-naive patients with schizophrenia. Am. J. Neuroradiol. 35, 30–37.

Folsom, D., Jeste, D.V., 2002. Schizophrenia in homeless persons: a systematic review of the literature. Acta Psychiatr. Scand. 105, 404–413.

Fox, M.D., Raichle, M.E., 2007. Spontaneous fluctuations in brain activity observed with functional magnetic resonance imaging. Nat. Rev. Neurosci. 8, 700–711.

Franke, B., Stein, J.L., Ripke, S., Anttila, V., Hibar, D.P., van Hulzen, K.J.E., Arias-Vasquez, A., Smoller, J.W., Nichols, T.E., Neale, M.C., McIntosh, A.M., Lee, P., McMahon, F.J., Meyer-Lindenberg, A., Mattheisen, M., Andreassen, O.A., Gruber, O., Sachdev, P.S., Roiz-Santiañez, R., Saykin, A.J., Ehrlich, S., Mather, K.A., Turner, J.A., Schwarz, E., Thalamuthu, A., Shugart, Y.Y., Ho, Y.Y., Martin, N.G., Wright, M.J., O'Donovan, M.C., Thompson, P.M., Neale, B.M., Medland, S.E., Sullivan, P.F., 2016. Genetic influences on schizophrenia and subcortical brain volumes: large-scale proof-of-concept and roadmap for future studies. Nat. Neurosci. 19, 420–431.

Fried, I., Rutishauser, U., Cerf, M., Kreiman, G., 2014. Single Neuron Studies of the Human Brain: Probing Cognition. MIT Press, Cambridge, MA; London, England.

Friston, K.J., Frith, C.D., 1995. Schizophrenia: a disconnection syndrome. Clin. Neurosci. 3, 89–97.

Friston, K.J., Rotshtein, P., Geng, J.J., Sterzer, P., Henson, R.N., 2006. A critique of functional localisers. Neuroimage 30, 1077–1087.

Garrity, A.G., Pearlson, G.D., McKiernan, K., Lloyd, D., Kiehl, K.A., Calhoun, V.D., 2007. Aberrant 'default mode' functional connectivity in schizophrenia. Am. J. Psychiatry 164, 450–457.

Grace, A.A., 1991. Phasic versus tonic dopamine release and the modulation of dopamine system responsivity: a hypothesis for the etiology of schizophrenia. Neuroscience 41, 1–24.

Grace, A.A., 2012. Dopamine system dysregulation by the hippocampus: implications for the pathophysiology and treatment of schizophrenia. Neuropharmacology 62, 1342–1348.

Hagmann, P., Jonasson, L., Maeder, P., Thiran, J.-P., Wedeen, V.J., Meuli, R., 2006. Understanding diffusion MR imaging techniques: from scalar diffusion-weighted imaging to diffusion tensor imaging and beyond. Radiographics 26, S205–S223.

Hall, J., Trent, S., Thomas, K.L., O'Donovan, M.C., Owen, M.J., 2015. Genetic risk for schizophrenia: convergence on synaptic pathways involved in plasticity. Biol. Psychiatry 77, 52–58. (The New Psychiatric Genetics: Toward Next Generation Diagnosis and Treatment).

Hallquist, M.N., Hillary, F.G., 2018. Graph Theory Approaches to Functional Network Organization in Brain Disorders: A Critique for a Brave New Small-World. bioRxiv. 012401.

Hawton, K., van Heeringen, K., 2009. Suicide. Lancet 373, 1372–1381.

Hindriks, R., Adhikari, M.H., Murayama, Y., Ganzetti, M., Mantini, D., Logothetis, N.K., Deco, G., 2016. Can sliding-window correlations reveal dynamic functional connectivity in resting-state fMRI? Neuroimage 127, 242–256.

Hoenig, J., 1983. The concept of schizophrenia. Kraepelin-Bleuler-Schneider. Br. J. Psychiatry 142, 547–556.

Honey, C.J., Sporns, O., Cammoun, L., Gigandet, X., Thiran, J.-P., Meuli, R., Hagmann, P., 2009. Predicting human resting-state functional connectivity from structural connectivity. Proc. Natl. Acad. Sci. 106, 2035–2040.

Hor, K., Taylor, M., 2010. Review: suicide and schizophrenia: a systematic review of rates and risk factors. J. Psychopharmacol. (Oxf.) 24, 81–90.

Hutchison, R.M., Womelsdorf, T., Gati, J.S., Everling, S., Menon, R.S., 2013. Resting-state networks show dynamic functional connectivity in awake humans and anesthetized macaques. Hum. Brain Mapp. 34, 2154–2177.

Insel, T., Cuthbert, B., Garvey, M., Heinssen, R., Pine, D.S., Quinn, K., Sanislow, C., Wang, P., 2010. Research domain criteria (RDoC): toward a new classification framework for research on mental disorders. Am. Psychiatr. Assoc. 748–751.

Jafri, M.J., Pearlson, G.D., Stevens, M., Calhoun, V.D., 2008. A method for functional network connectivity among spatially independent resting-state components in schizophrenia. Neuroimage 39, 1666–1681.

Jones, D.K., Leemans, A., 2011. Diffusion tensor imaging. In: Magnetic Resonance Neuroimaging, Methods in Molecular Biology. Humana Press, pp. 127–144.

Kambeitz, J., Kambeitz-Ilankovic, L., Cabral, C., Dwyer, D.B., Calhoun, V.D., Heuvel, V.D., P, M., Falkai, P., Koutsouleris, N., Malchow, B., 2016. Aberrant functional whole-brain network architecture in patients with schizophrenia: a meta-analysis. Schizophr. Bull. 42, S13–S21.

Kanaan, R.A., Picchioni, M.M., McDonald, C., Shergill, S.S., McGuire, P.K., 2017. White matter deficits in schizophrenia are global and don't progress with age. Aust. N. Z. J. Psychiatry 51, 1020–1031.

Karlsgodt, K.H., van Erp, T.G.M., Poldrack, R.A., Bearden, C.E., Nuechterlein, K.H., Cannon, T.D., 2008. Diffusion tensor imaging of the superior longitudinal fasciculus and working memory in recent-onset schizophrenia. Biol. Psychiatry 63, 512–518.

Karow, A., Wittmann, L., Schöttle, D., Schäfer, I., Lambert, M., 2014. The assessment of quality of life in clinical practice in patients with schizophrenia. Dialogues Clin. Neurosci. 16, 185–195.

Kay, S.R., Flszbein, A., Opfer, L.A., 1987. The positive and negative syndrome scale (PANSS) for schizophrenia. Schizophr. Bull. 13, 261.

Kelleher, I., Corcoran, P., Keeley, H., Wigman, J.T.W., Devlin, N., Ramsay, H., Wasserman, C., Carli, V., Sarchiapone, M., Hoven, C., Wasserman, D., Cannon, M., 2013. Psychotic symptoms and population risk for suicide attempt: a prospective cohort study. JAMA Psychiat. 70, 940–948.

Keshavan, M.S., Lawler, A.N., Nasrallah, H.A., Tandon, R., 2017. New drug developments in psychosis: challenges, opportunities and strategies. Prog. Neurobiol. 152, 3–20. Developing drugs for neurological and psychiatric disorders: challenges and opportunities.

Khanna, A., Pascual-Leone, A., Michel, C.M., Farzan, F., 2015. Microstates in resting-state EEG: current status and future directions. Neurosci. Biobehav. Rev. 49, 105–113.

Kim, D.-J., Kent, J.S., Bolbecker, A.R., Sporns, O., Cheng, H., Newman, S.D., Puce, A., O'Donnell, B.F., Hetrick, W.P., 2014. Disrupted modular architecture of cerebellum in schizophrenia: a graph theoretic analysis. Schizophr. Bull. 40, 1216–1226.

Klingner, C.M., Langbein, K., Dietzek, M., Smesny, S., Witte, O.W., Sauer, H., Nenadic, I., 2014. Thalamocortical connectivity during resting state in schizophrenia. Eur. Arch. Psychiatry Clin. Neurosci. 264, 111–119.

Koch, M.A., Norris, D.G., Hund-Georgiadis, M., 2002. An Investigation of Functional and Anatomical Connectivity Using Magnetic Resonance Imaging. Neuroimage 16, 241–250.

Kotov, R., Krueger, R.F., Watson, D., 2018. A paradigm shift in psychiatric classification: the hierarchical taxonomy of psychopathology (HiTOP). World Psychiat. 17, 24–25.

Kraepelin, E., 1899. Psychiatrie: Ein Lehrbuch fuer Studirende und Aerzte, sixth ed. J. A. Barth, Leipzig.

Kusumi, I., Boku, S., Takahashi, Y., 2015. Psychopharmacology of atypical antipsychotic drugs: from the receptor binding profile to neuroprotection and neurogenesis. Psychiatry Clin. Neurosci. 69, 243–258.

Lally, J., MacCabe, J.H., 2015. Antipsychotic medication in schizophrenia: a review. Br. Med. Bull. 114, 169–179.

Laursen, T.M., Nordentoft, M., Mortensen, P.B., 2014. Excess early mortality in schizophrenia. Annu. Rev. Clin. Psychol. 10, 425–448.

Lawrie, S.M., Buechel, C., Whalley, H.C., Frith, C.D., Friston, K.J., Johnstone, E.C., 2002. Reduced frontotemporal functional connectivity in schizophrenia associated with auditory hallucinations. Biol. Psychiatry 51, 1008–1011.

Liang, M., Zhou, Y., Jiang, T., Liu, Z., Tian, L., Liu, H., Hao, Y., 2006. Widespread functional disconnectivity in schizophrenia with resting-state functional magnetic resonance imaging. Neuroreport 17, 209–213.

Lindquist, M.A., Xu, Y., Nebel, M.B., Caffo, B.S., 2014. Evaluating dynamic bivariate correlations in resting-state fMRI: a comparison study and a new approach. Neuroimage 101, 531–546.

Lynall, M.-E., Bassett, D.S., Kerwin, R., McKenna, P.J., Kitzbichler, M., Muller, U., Bullmore, E., 2010. Functional connectivity and brain networks in schizophrenia. J. Neurosci. 30, 9477–9487.

Makeig, S., Debener, S., Onton, J., Delorme, A., 2004. Mining event-related brain dynamics. Trends Cogn. Sci. 8, 204–210.

Matsui, T., Murakami, T., Ohki, K., 2018. Neuronal origin of the temporal dynamics of spontaneous BOLD activity correlation. Cerebral Cortex. (in press).

McGlashan, T., Walsh, B., Woods, S., 2010. The Psychosis-Risk Syndrome: Handbook for Diagnosis and Follow-Up. Oxford University Press, New York.

McGorry, P.D., Nelson, B., Amminger, G.P., Bechdolf, A., Francey, S.M., Berger, G., Riecher-Rössler, Klosterkötter, J., Ruhrmann, S., Schultze-Lutter, F., 2009. Intervention in individuals at ultra high risk for psychosis: a review and future directions. J. Clin. Psychiatry 70, 1206–1212.

McKeown, M.J., Makeig, S., Brown, G.G., Jung, T.-P., Kindermann, S.S., Bell, A.J., Sejnowski, T.J., 1998. Analysis of fMRI data by blind separation into independent spatial components. Hum. Brain Mapp. 6, 160–188.

Merricks, E.M., Smith, E.H., McKhann, G.M., Goodman, R.R., Bateman, L.M., Emerson, R.G., Schevon, C.A., Trevelyan, A.J., 2015. Single unit action potentials in humans and the effect of seizure activity. Brain 138, 2891–2906.

Miller, R.L., Yaesoubi, M., Turner, J.A., Mathalon, D.H., Preda, A., Pearlson, G., Adali, T., Calhoun, V.D., 2016. Higher dimensional meta-state analysis reveals reduced resting fMRI connectivity dynamism in schizophrenia patients. PLoS One 11, e0149849.

Miller, R.L., Adali, T., Levin-Schwartz, Y., Calhoun, V.D., 2017. Resting-State fMRI Dynamics and Null Models: Perspectives, Sampling Variability, and Simulations. bioRxiv. 153411.

Mori, S., van Zijl, P.C.M., 2002. Fiber tracking: principles and strategies—a technical review. NMR Biomed. 15, 468–480.

Nielsen, R.E., Uggerby, A.S., Jensen, S.O.W., McGrath, J.J., 2013. Increasing mortality gap for patients diagnosed with schizophrenia over the last three decades— a Danish nationwide study from 1980 to 2010. Schizophr. Res. 146, 22–27.

O'Donovan, M.C., 2015. What have we learned from the psychiatric genomics consortium. World Psychiat. 14, 291–293.

Olfson, M., Gerhard, T., Huang, C., Crystal, S., Stroup, T.S., 2015. Premature mortality among adults with schizophrenia in the United States. JAMA Psychiat. 72, 1172–1181.

Ombao, H., Bellegem, S.V., 2008. Evolutionary coherence of nonstationary signals. IEEE Trans. Signal Proc. 56, 2259–2266.

Owen, P.R., 2012. Portrayals of schizophrenia by entertainment media: a content analysis of contemporary movies. Psychiatr. Serv. 63, 655–659.

Poels, E.M.P., Kegeles, L.S., Kantrowitz, J.T., Slifstein, M., Javitt, D.C., Lieberman, J.A., Abi-Dargham, A., Girgis, R.R., 2014. Imaging glutamate in schizophrenia: review of findings and implications for drug discovery. Mol. Psychiatry 19, 20–29.

Poldrack, R.A., 2007. Region of interest analysis for fMRI. Soc. Cogn. Affect. Neurosci. 2, 67–70.

Power, J.D., Mitra, A., Laumann, T.O., Snyder, A.Z., Schlaggar, B.L., Petersen, S.E., 2014. Methods to detect, characterize, and remove motion artifact in resting state fMRI. Neuroimage 84, 320–341.

Raichle, M.E., 2015. The Brain's default mode network. Annu. Rev. Neurosci. 38, 433–447.

Raichle, M.E., MacLeod, A.M., Snyder, A.Z., Powers, W.J., Gusnard, D.A., Shulman, G.L., 2001. A default mode of brain function. Proc. Natl. Acad. Sci. 98, 676–682.

Rashid, B., Damaraju, E., Pearlson, G.D., Calhoun, V.D., 2014. Dynamic connectivity states estimated from resting fMRI identify differences among schizophrenia, bipolar disorder, and healthy control subjects. Front. Hum. Neurosci. 8, 897.

Rashid, B., Arbabshirani, M.R., Damaraju, E., Cetin, M.S., Miller, R., Pearlson, G.D., Calhoun, V.D., 2016. Classification of schizophrenia and bipolar patients using static and dynamic resting-state fMRI brain connectivity. Neuroimage 134, 645–657.

Rubinov, M., Sporns, O., 2010. Complex network measures of brain connectivity: Uses and interpretations. Neuroimage 52, 1059–1069. Computational Models of the Brain.

Ruhrmann, S., Schultze-Lutter, F., Schmidt, S.J., Kaiser, N., Klosterkötter, J., 2014. Prediction and prevention of psychosis: current progress and future tasks. Eur. Arch. Psychiatry Clin. Neurosci. 264, 9–16.

Sakoğlu, Ü., Pearlson, G.D., Kiehl, K.A., Wang, Y.M., Michael, A.M., Calhoun, V.D., 2010. A method for evaluating dynamic functional network connectivity and task-modulation: application to schizophrenia. Magn. Reson. Mater. Phys. Biol. Med. 23, 351–366.

Samartzis, L., Dima, D., Fusar-Poli, P., Kyriakopoulos, M., 2014. White matter alterations in early stages of schizophrenia: a systematic review of diffusion tensor imaging studies: white matter alterations in early schizophrenia. J. Neuroimaging 24, 101–110.

Sato, M., 2006. Renaming schizophrenia: a Japanese perspective. World Psychiat. 5, 53–55.

Schmitt, A., Malchow, B., Hasan, A., Fallkai, P., 2014. The impact of environmental factors in severe psychiatric disorders. Front. Neurosci. 8, 19.

Senkowski, D., Gallinat, J., 2015. Dysfunctional prefrontal gamma-band oscillations reflect working memory and other cognitive deficits in schizophrenia. Biol. Psychiatry 77, 1010–1019. Cortical Oscillations for Cognitive/Circuit Dysfunction in Psychiatric Disorders.

Shakil, S., Lee, C.-H., Keilholz, S.D., 2016. Evaluation of sliding window correlation performance for characterizing dynamic functional connectivity and brain states. Neuroimage 133, 111–128.

Shehzad, Z., Kelly, A.C., Reiss, P.T., Gee, D.G., Gotimer, K., Uddin, L.Q., Lee, S.H., Margulies, D.S., Roy, A.K., Biswal, B.B., Petkova, E., 2009. The resting brain: Unconstrained yet reliable. Cereb. Cortex 19, 2209–2229.

Shen, H., Wang, L., Liu, Y., Hu, D., 2010. Discriminative analysis of resting-state functional connectivity patterns of schizophrenia using low dimensional embedding of fMRI. Neuroimage 49, 3110–3121.

Shine, J.M., Bissett, P.G., Bell, P.T., Koyejo, O., Balsters, J.H., Gorgolewski, K.J., Moodie, C.A., Poldrack, R.A., 2016. The dynamics of functional brain networks: integrated network states during cognitive task performance. Neuron 92, 544–554.

Shinn, A.K., Baker, J.T., Lewandowski, K.E., Öngür, D., Cohen, B.M., 2015. Aberrant cerebellar connectivity in motor and association networks in schizophrenia. Front. Hum. Neurosci. 9, 134.

Shorter, K.R., Miller, B.H., 2015. Epigenetic mechanisms in schizophrenia. Prog. Biophys. Mol. Biol. 118, 1–7. Epigenetic Inheritance and Programming.

Simeone, J.C., Ward, A.J., Rotella, P., Collins, J., Windisch, R., 2015. An evaluation of variation in published estimates of schizophrenia prevalence from 1990–2013: a systematic literature review. BMC Psychiatry 15, 193.

Skudlarski, P., Jagannathan, K., Calhoun, V.D., Hampson, M., Skudlarska, B.A., Pearlson, G., 2008. Measuring brain connectivity: diffusion tensor imaging validates resting state temporal correlations. Neuroimage 43, 554–561.

Smith, S.M., Vidaurre, D., Beckmann, C.F., Glasser, M.F., Jenkinson, M., Miller, K.L., Nichols, T.E., Robinson, E.C., Salimi-Khorshidi, G., Woolrich, M.W., Barch, D.M., Uğurbil, K., Van Essen, D.C., 2013. Functional connectomics from resting-state fMRI. Trends Cogn. Sci. 17, 666–682. Special Issue: The Connectome.

Soares, J.M., Marques, P., Alves, V., Sousa, N., 2013. A hitchhiker's guide to diffusion tensor imaging. Front. Neurosci. 7, 31.

Steinley, D., 2006. K-means clustering: a half-century synthesis. Br. J. Math. Stat. Psychol. 59, 1–34.

Sui, J., He, H., Yu, Q., Chen, J., Rogers, J., Pearlson, G.D., Mayer, A., Bustillo, J., Canive, J., Calhoun, V.D., 2013. Combination of resting state fMRI, DTI, and sMRI data to discriminate schizophrenia by N-way MCCA + jICA. Front. Hum. Neurosci. 7, 235.

Sui, J., Pearlson, G.D., Du, Y., Yu, Q., Jones, T.R., Chen, J., Jiang, T., Bustillo, J., Calhoun, V.D., 2015. In search of multimodal neuroimaging biomarkers of cognitive deficits in schizophrenia. Biol. Psychiatry 78, 794–804. Schizophrenia: Glutamatergic Mechanisms of Cognitive Dysfunction and Treatment.

The Network and Pathway Analysis Subgroup of the Psychiatric Genomics Consortium, 2015. Psychiatric genome-wide association study analyses implicate neuronal, immune and histone pathways. Nat. Neurosci. 18, 199–209.

Tzourio-Mazoyer, N., Landeau, B., Papathanassiou, D., Crivello, F., Etard, O., Delcroix, N., Mazoyer, B., Joliot, M., 2002. Automated anatomical labeling of activations in SPM using a macroscopic anatomical parcellation of the MNI MRI single-subject brain. Neuroimage 15, 273–289.

van den Heuvel, M.P., Fornito, A., 2014. Brain networks in schizophrenia. Neuropsychol. Rev. 24, 32–48.

van den Heuvel, M.P., Sporns, O., 2011. Rich-club organization of the human connectome. J. Neurosci. 31, 15775–15786.

van den Heuvel, M.P., Sporns, O., Collin, G., Scheewe, T., Mandl, R.C.W., Cahn, W., Goñi, J., Hulshoff Pol, H.E., Kahn, R.S., 2013. Abnormal Rich Club organization and functional brain dynamics in schizophrenia. JAMA Psychiat. 70, 783.

Van Os, J., 2009. 'Salience syndrome' replaces 'schizophrenia' in DSM-V and ICD-11: psychiatry's evidence-based entry into the 21st century? Acta Psychiatr. Scand. 120, 363–372.

van Os, J., Guloksuz, S., 2017. A critique of the 'ultra-high risk' and 'transition' paradigm. World Psychiat. 16, 200–206.

Varoquaux, G., Craddock, R.C., 2013. Learning and comparing functional connectomes across subjects. Neuroimage 80, 405–415. Mapping the Connectome.

Vatansever, D., Menon, D.K., Manktelow, A.E., Sahakian, B.J., Stamatakis, E.A., 2015. Default mode dynamics for global functional integration. J. Neurosci. 35, 15254–15262.

Voineskos, A.N., Lobaugh, N.J., Bouix, S., Rajji, T.K., Miranda, D., Kennedy, J.L., Mulsant, B.H., Pollock, B.G., Shenton, M.E., 2010. Diffusion tensor tractography findings in schizophrenia across the adult lifespan. Brain 133, 1494–1504.

Wang, Q., Su, T.-P., Zhou, Y., Chou, K.-H., Chen, I.-Y., Jiang, T., Lin, C.-P., 2012. Anatomical insights into disrupted small-world networks in schizophrenia. Neuroimage 59, 1085–1093.

Weinstein, J.J., van de Giessen, E., Rosengard, R.J., Xu, X., Ojeil, N., Brucato, G., Gil, R.B., Kegeles, L.S., Laruelle, M., Slifstein, M., Abi-Dargham, A., 2017. PET imaging of dopamine-D2 receptor internalization in schizophrenia. Mol. Psychiatry.

Whitfield-Gabrieli, S., Ford, J.M., 2012. Default mode network activity and connectivity in psychopathology. Annu. Rev. Clin. Psychol. 8, 49–76.

Whitfield-Gabrieli, S., Thermenos, H.W., Milanovic, S., Tsuang, M.T., Faraone, S.V., McCarley, R.W., Shenton, M.E., Green, A.I., Nieto-Castanon, A., LaViolette, P., 2009. Hyperactivity and hyperconnectivity of the default network in schizophrenia and in first-degree relatives of persons with schizophrenia. Proc. Natl. Acad. Sci. 106, 1279–1284.

Woodward, N.D., Karbasforoushan, H., Heckers, S., 2012. Thalamocortical dysconnectivity in schizophrenia. Am. J. Psychiatry 169, 1092–1099.

Wotruba, D., Michels, L., Buechler, R., Metzler, S., Theodoridou, A., Gerstenberg, M., Walitza, S., Kollias, S., Rössler, W., Heekeren, K., 2013. Aberrant coupling within and across the default mode, task-positive, and salience network in subjects at risk for psychosis. Schizophr. Bull. 40, 1095–1104.

Wu, C.-H., Hwang, T.-J., Chen, Y.-J., Hsu, Y.-C., Lo, Y.-C., Liu, C.-M., Hwu, H.-G., Liu, C.-C., Hsieh, M.H., Chien, Y.L., Chen, C.-M., Isaac Tseng, W.-Y., 2015. Altered integrity of the right arcuate fasciculus as a trait marker of schizophrenia: a sibling study using tractography-based analysis of the whole brain. Hum. Brain Mapp. 36, 1065–1076.

Yaesoubi, M., Allen, E.A., Miller, R.L., Calhoun, V.D., 2015. Dynamic coherence analysis of resting fMRI data to jointly capture state-based phase, frequency, and time-domain information. Neuroimage 120, 133–142.

Yaesoubi, M., Adalı, T., Calhoun, V.D., 2018. A window-less approach for capturing time-varying connectivity in fMRI data reveals the presence of states with variable rates of change. Hum. Brain Mapp. 39 (4), 1626–1636.

Yeiser, B., 2017. My triumph over psychosis: a journey from schizophrenia and homelessness to college graduate. Schizophr. Bull. 43, 943–945.

Yu, Q., Sui, J., Rachakonda, S., He, H., Gruner, W., Pearlson, G., Kiehl, K.A., Calhoun, V.D., 2011. Altered topological properties of functional network connectivity in schizophrenia during resting state: A small-world brain network study. PLoS One 6, e25423.

Yu, Q., Plis, S.M., Erhardt, E.B., Allen, E.A., Sui, J., Kiehl, K.A., Pearlson, G., Calhoun, V.D., 2012. Modular organization of functional network connectivity in healthy controls and patients with schizophrenia during the resting state. Front. Syst. Neurosci. 5, 103.

Yu, Q., Sui, J., Liu, J., Plis, S.M., Kiehl, K.A., Pearlson, G., Calhoun, V.D., 2013. Disrupted correlation between low frequency power and connectivity strength of resting state brain networks in schizophrenia. Schizophr. Res. 143, 165–171.

Yu, Q., Erhardt, E.B., Sui, J., Du, Y., He, H., Hjelm, D., Cetin, M.S., Rachakonda, S., Miller, R.L., Pearlson, G., Calhoun, V.D., 2015. Assessing dynamic brain graphs of time-varying connectivity in fMRI data: application to healthy controls and patients with schizophrenia. Neuroimage 107, 345–355.

Yu, Q., Du, Y., Chen, J., He, H., Sui, J., Pearlson, G., Calhoun, V.D., 2017. Comparing brain graphs in which nodes are regions of interest or independent components: A simulation study. J. Neurosci. Methods 291, 61–68.

Zalesky, A., Breakspear, M., 2015. Towards a statistical test for functional connectivity dynamics. Neuroimage 114, 466–470.

Zou, Q.-H., Zhu, C.-Z., Yang, Y., Zuo, X.-N., Long, X.-Y., Cao, Q.-J., Wang, Y.-F., Zang, Y.-F., 2008. An improved approach to detection of amplitude of low-frequency fluctuation (ALFF) for resting-state fMRI: fractional ALFF. J. Neurosci. Methods 172, 137–141.

Genetics of brain networks and connectivity

Emily L. Dennis, Paul M. Thompson, Neda Jahanshad

Imaging Genetics Center, Mark and Mary Stevens Neuroimaging and Informatics Institute, Keck School of Medicine of the University of Southern California, Marina del Rey, CA, United States

CHAPTER OUTLINE

MOTIVATION FOR GENETIC STUDIES OF THE BRAIN'S STRUCTURAL AND FUNCTIONAL CONNECTIVITY

The quest to identify genetic factors that affect the brain has several motivations. From a neuroscientific standpoint, there is a broad range of individual variation in all human traits including all aspects of behavior and cognition; population studies have

Connectomics. https://doi.org/10.1016/B978-0-12-813838-0.00008-X

shown these traits are partially influenced by variations in each individual's genetic code, among many other factors such as environmental hazards, toxins, trauma, nutrition, and education. With the advent of gene sequencing technology—and more recently the capacity to sequence an individual's genome—the scope of genetic studies has greatly advanced to include studies that try to pinpoint individual loci in the genome, or sets of such loci, where variations have a measurable impact on the human brain and behavior. At the same time, the whole field of genetics in animal models—from invertebrates to transgenic and outbred mice—has allowed, in some cases, direct manipulation of the genetic code to identify specific causal links between genetic variation, the brain, and behavior.

In this chapter, we review some of the recent efforts to relate genetic variation to variations in the human brain, with a focus on measures of brain connectivity, given the scope of this book. The field of imaging genetics is a vast and ever growing activity. Large international projects such as ENIGMA, CHARGE, and national biobanks, are fueling the development of mathematical and "big data" analytic methods to relate genetic variation to brain variation in populations across the world.

There are several motivations for finding genetic loci and pathways that affect brain variation and connectivity. We know from work in *Aplysia* and *C. elegans* that there are intricate genetic programs that guide cell proliferation, axon guidance, and synaptic connectivity in simpler invertebrate nervous systems. These genetic programs involve some of the same molecular pathways that are altered in humans. In monogenic neurodevelopmental disorders such as Fragile X syndrome, Williams syndrome, and 22q deletion syndrome, gene sequencing has identified consistent genetic deletions or expansions that cause specific conditions, and connectomics studies are charting their effects on the brain in ever increasing detail. Large scale international efforts are now relating the size and scope of these genetic deletions and duplications to alterations in brain connectivity and microstructure (Lin et al., 2017; Sonderby et al., 2017). Much of the genetic risk for human brain disease and neuropsychiatric illness is mediated by common genetic variation—variants at single loci in the genome that collectively alter a person's risk for disorders as diverse as Alzheimer's disease, schizophrenia, PTSD, and autism. In these studies, neuroimaging can offer an "endophenotype" or intermediate phenotype to facilitate the genetic study of specific brain systems and circuits. When brain images are subjected to genetic analysis in large scale international studies, there is often a focus on brain systems that might mediate disease risk, to better understand treatment targets and potential modulators of disease risk and progression.

TO WHAT EXTENT ARE BRAIN VARIATIONS INFLUENCED BY GENETICS? A HISTORY OF HERITABILITY WITH TWIN AND FAMILY STUDIES

Twin and family studies have revealed that the structural connectivity of the brain is under strong genetic control (Kochunov et al., 2010). Chiang et al. (2009) found that genetic factors accounted for 55%–85% of the variance in fractional anisotropy

(FA)—a widely used measure of white matter microstructure—across the frontal, parietal, and occipital lobes. Similar estimates have been shown in neonatal twin studies using both an ROI approach (Geng et al., 2012) and tractography (Lee et al., 2015). A megaanalysis of two large family studies showed a high level of heritability across the WM, with central WM showing greater heritability generally than peripheral WM (Neda et al., 2013a). Similar patterns were shown in a twin study of middle-aged men (Vuoksimaa et al., 2017). Exceptions to this pattern include the cortico-spinal tract and fornix, which yielded low heritability estimates in some analyses. This may be partially due to poorer reliability in reconstructing and measuring diffusivity in this tract. Focusing on tracts of the limbic system, (Budisavljevic et al., 2016) reported very high heritability for the FA of the uncinate, with moderate to high heritability for the cingulum and fornix. Even fiber orientation has a demonstrable genetic component (Shen et al., 2014). Examining left-right hemisphere asymmetry, Jahanshad et al. (2010) found that the heritability of asymmetry varied across tracts, with genetic components explaining the most variance in the asymmetry of the anterior thalamic radiation and inferior fronto-occipital fasciculus. In the near future, large-scale studies of genetic variations and brain structure (Hibar et al., 2017; Adams et al., 2016) will be adapted to study measures of hemispheric specialization and laterality, which can be measured on a scale where genome-wide searches are well powered and feasible (Guadalupe et al., 2017).

Functional connectivity also has a significant heritable component, although generally estimates are lower than they are for structural connectivity. In resting networks reconstructed from EEG, heritability estimates for measures of network clustering and path length range between 23% and 89%, depending on the frequency band (Smit et al., 2008; Schutte et al., 2013). This high heritability has motivated large scale genetic screens for EEG features. Connectivity patterns from MEG can even be used to identify a co-twin with 75% accuracy (Demuru et al., 2017). Using ICA on rsfMRI time series, (Glahn et al., 2010) found that a little less than half of the variance in default mode network connectivity was accounted for by genetic components. They also found that the genetic factors that influenced functional connectivity were distinct from those that influenced brain structure. (Yang et al., 2016) showed that this heritability holds for variation in both within network and between network connectivity. Across two large family cohorts, (Adhikari et al., 2018) used a seed-based approach to extract 20 functional networks, showing consistent heritability estimates between 20% and 40% across the networks. Using a graph theory approach in a small twin study, (Fornito et al., 2011) found that genetic components accounted for 60% of the variance in the cost-efficiency of the network. In a similar cohort, (van den Heuvel et al., 2013) reported that genetic components accounted for 42% of the variance in a global measure of normalized path length, closely related to network efficiency. A larger sample replicated these results, showing moderate estimates across measures, particularly when connection density in the network was restricted (Sinclair et al., 2015). An important consideration of studies of the genetics of functional connectivity is that head motion is moderately heritable, which has the potential to influence heritability estimates (Couvy-Duchesne et al., 2014).

ALTERED BRAIN CONNECTIVITY IN NEUROGENETIC DISORDERS AND GENETIC DELETIONS AND DUPLICATIONS

Mutations give researchers an opportunity to draw more precise conclusions about causal associations between specific genes and specific functions. Below we consider a number of neurogenetic disorders, including disease caused by a mutation in a single gene (Huntington's Disease, Fragile X), deletion syndromes (22q11.2 DS, Williams Syndrome, Prader-Willi Syndrome), and chromosomal abnormalities (Turner Syndrome, Down Syndrome). Alterations in both structural and functional connectivity are discussed.

HUNTINGTON'S DISEASE

Huntington's disease is caused by a mutation in the *HTT* gene (huntingin) which leads to excessive CAG repeats. The resulting protein is abnormally long, cleaved into fragments, which bind and accumulate in neurons, disrupting function and eventually causing neuronal death (MacDonald et al., 1993). Patients with HD experience a progressive loss of motor control including characteristic twitching movements (chorea), along with emotional and cognitive disruptions. Disruptions to structural and functional connectivity can be seen both before and after onset of the disease. Premanifest carriers of the genetic mutation, at the group level, show lower FA, higher MD and RD in the corpus callosum, posterior internal capsule, caudate, and the premotor and motor cortices (Rosas et al., 2010; Novak et al., 2014; Dumas et al., 2012; Bohanna et al., 2011a; Shaffer et al., 2017; Matsui et al., 2015; Phillips et al., 2013, 2016; Poudel et al., 2014b; Di Paola et al., 2012), along with higher FA in the anterior internal capsule and putamen (Rosas et al., 2006; Magnotta et al., 2009). Patients in the early phase of the disease showed more extensive differences in these areas (Rosas et al., 2006, 2010; Novak et al., 2014; Dumas et al., 2012; Phillips et al., 2013; Poudel et al., 2014b; Di Paola et al., 2012). Patients with reduced functional capacity or a greater number of CAG repeats showed more widespread and pronounced disruptions (Rosas et al., 2010; Matsui et al., 2015, 2014; Phillips et al., 2013, 2016). Longitudinal studies show these alterations to be progressive (Weaver et al., 2009; Poudel et al., 2015; Gregory et al., 2015; Harrington et al., 2016; Shaffer et al., 2017; Domínguez et al., 2013). Bohanna et al. (2011a,b) showed that WM deficits in the body of the corpus callosum were associated with motor disruption, while deficits in the genu of the corpus callosum were associated with cognitive impairment. Focusing on fibers connecting the basal ganglia to the cortex, Novak et al. (2015) found largely intact tracts in premanifest HD patients, while patients in the early phase of the disease did show significant alterations. Prior to onset of HD, patients exhibit reduced connectivity between the frontal and motor cortex, along with the striatum (Dumas et al., 2013; Poudel et al., 2014a; Unschuld et al., 2012; Wolf et al., 2008). These disruptions become more widespread in the early stage of the disease, including the basal ganglia and additional cortical regions (Wolf et al., 2014). After HD onset, functional connectivity disruptions are extensive, encompassing the

DMN, ECN, and DAN (Dumas et al., 2013; Quarantelli et al., 2013; Poudel et al., 2014a; Thiruvady et al., 2007). Greater disease burden, caused by a larger number of CAG repeats, is associated with more pronounced disruptions in connectivity (Koenig et al., 2014; Espinoza et al., 2018), which is further associated with more pronounced cognitive and motor disruption (Espinoza et al., 2018). Using a graph theory approach, studies have argued that there may be a transition toward more "random" networks as disease severity increases (Harrington et al., 2015; Gargouri et al., 2016). Integrating structural and functional connectivity measures, McColgan et al. (2017) found both increases and decreases in connectivity, generally following an anterior-posterior gradient, suggesting compensatory connectivity in anterior regions as posterior regions begin to degenerate.

FRAGILE X SYNDROME

Fragile X (FX) syndrome results from the expansion of a CGG repeat in the 5′ untranslated region of the fragile X mental retardation 1 gene (*FMR1*). This causes a loss or disruption to the functionality of the associated protein, fragile X mental retardation protein (FMRP) (Hagerman et al., 2017). It is a common cause of intellectual disability, especially in boys (Verkerk et al., 1991). Alterations in frontostriatal and thalamocortical pathways have been shown in structural connectivity studies of FX, primarily in the form of lower FA (Barnea-Goraly et al., 2003a,b; Haas et al., 2009; Villalon-Reina et al., 2013). Increases in FA have also been reported in some association fibers (Hall et al., 2016). Using a graph theory approach, Bruno et al. (2017) found that the balance of integration and segregation in the structural connectome was disturbed, primarily due to decreased local segregation. There have not been many studies of functional connectivity in FX. Hall et al. (2013) reported widespread reductions in functional connectivity in FX, including the salience, executive control, and visuospatial networks. They further found that within FX patients, connectivity of the left insula was positively correlated with IQ.

22Q11.2 DELETION SYNDROME

22q11.2 Deletion Syndrome (22q DS), also called velocardiofacial syndrome or DiGeorge syndrome, is caused by a deletion on chromosome 22. 22q DS is associated with a range of physiological, neurological, and psychological symptoms (Shprintzen, 2000), including a greatly increased risk of developing schizophrenia (Bassett and Chow, 1999). 22q DS is associated with both increases and decreases in FA. Increased FA has been reported in the cingulum, frontal, and parietal lobe (Simon et al., 2005, 2008), while decreased FA has been reported more widely, including the corpus callosum and association fibers (Barnea-Goraly et al., 2003a,b; Villalon-Reina et al., 2013; Jalbrzikowski et al., 2014; Olszewski et al., 2017). The cingulum similarly shows a mixture of higher and lower FA in 22q DS (Kates et al., 2015; Nuninga et al., 2017). The degree of these alterations is further associated with the severity of cognitive and psychotic symptoms (Jalbrzikowski et al., 2014; Olszewski et al., 2017), and longitudinal analysis shows a lack of age-expected changes in structural and functional connectivity (Jalbrzikowski et al., 2014; Padula et al., 2015). Examining functional connectivity using resting state fMRI (rsfMRI), several studies have shown altered connectivity, both increases and decreases, particularly among central nodes of the DMN (Padula et al., 2015, 2017; Schreiner et al., 2014; Debbané et al., 2012; Mattiaccio et al., 2016, 2018), which are similarly associated with symptom severity. One explanation for the mixture of both increased and decreased connectivity considers the schizophrenia research in this area—increased connectivity may be partially to blame for positive symptoms, such as hallucinations, while decreased connectivity may be partially to blame for the negative symptoms, such as poor social competence. Given the significant association between 22q DS and schizophrenia (up to ⅓ of patients develop psychosis), these may represent one mechanism through which psychological disturbances emerge.

WILLIAMS SYNDROME

Williams Syndrome (WS) is caused by a hemizygous deletion on chromosome 7 at locus 7q11.23 and leads to physiological, intellectual, and behavioral abnormalities (Bellugi et al., 2001). Structural connectivity studies have found both increases and decreases in FA in WS. Higher FA in WS has been reported in the superior and inferior longitudinal fasciculi (Hoeft et al., 2007; Arlinghaus et al., 2011; Haas et al., 2014), association pathways central to language, memory, and visual processing (Catani and de Schotten, 2012). Potentially conflicting results have been reported in the uncinate fasciculus, which is critical for emotion processing and social cognition. Both higher FA in children with WS (Haas et al., 2014) and lower FA in adults with WS (Jabbi et al., 2012; Avery et al. 2012) have been reported. RsfMRI studies of WS show decreased within-network connectivity, especially within the DMN (Vega et al., 2015; Sampaio et al., 2016).

PRADER-WILLI SYNDROME

Prader-Willi Syndrome (PWS) is caused by a loss of function on chromosome 15, either from a deletion in the paternal copy (q11–q13), or transmission of two maternal copies, one of which is partially silenced (Cassidy 1997). It is associated with intellectual disability, obesity, and behavioral problems. As it is less common than other neurogenetic conditions covered here, there is little brain imaging research on PWS. Several studies have reported lower FA in central WM tracts (Yamada et al., 2006; Rice et al., 2017; Lukoshe et al., 2017), as well as association tracts such as the IFOF, ILF, and SLF (Xu et al., 2017; Lukoshe et al., 2017; Rice et al., 2017). These alterations may underlie deficits in attention, language and sensorimotor processing, and emotion. Focusing on functional connectivity, (Pujol et al., 2016) showed abnormal connectivity with the basal ganglia, with specific alterations related to subtypes of OCD behavior. Patients with PWS may overeat, causing obesity. Alterations in circuits related to cognitive control and satiety are thought to underlie this, and functional connectivity studies focusing on prefrontal and hypothalamic circuitry have shown alterations (Zhang et al., 2013, 2015).

TURNER SYNDROME

Turner Syndrome (TS) results from the absence of one X chromosome in girls. It is associated with physical, hormonal, and neurological alterations. Structural connectivity studies of TS generally show lower FA across a large amount of the WM (Molko et al., 2004; Holzapfel et al., 2006; Yamagata et al., 2012; Villalon-Reina et al., 2013). Functional connectivity studies have shown reduced connectivity, which was associated with cognitive performance (Bray et al., 2011; Xie et al., 2017). Another analysis of this dataset revealed that subdivisions of the posterior parietal cortex are recruited to different networks in TS (Bray et al., 2012).

DOWN SYNDROME

Down Syndrome (DS), also called trisomy 21, is caused by the presence of a third chromosome 21. It is one of the most common chromosomal abnormalities (0.1%) and the most common cause of intellectual disability (Nadel, 1999). Individuals with DS are at a greater risk of developing dementia, so, many imaging studies focus on adult patients. DS is associated with widespread reductions in FA, particularly in central WM tracts such as the corpus callosum, internal capsule, and brainstem (Fenoll et al., 2017; Gunbey et al., 2017). Decreased FA has also been reported in the frontal cortex, with greater deficits shown in DS patients with dementia compared to those without dementia (Powell et al., 2014). RsfMRI studies have shown a disruption in functional connectivity, with increased connectivity between networks (Anderson et al., 2013; Vega et al., 2015). Pujol et al. (2015) showed higher connectivity among ventral structures in DS, while a network of dorsal executive structures showed lower connectivity.

DIGGING DEEPER—SEARCHING FOR THE EFFECT OF SINGLE NUCLEOTIDE POLYMORPHISMS

For the brain disorders described above, there is a genetic deletion or duplication that is known to cause the disorder, but the vast majority of psychiatric and neurological disorders in the general population are *complex traits*, and there is no single genetic test that unequivocally rules in or out the illness. Instead, a large number of variants in the genome—some common and some rare—influence disease risk, with none of them being sufficient for developing the disease. These variations in the genome occur at single nucleotide polymorphisms, (SNPs).

In the case of Alzheimer's disease, variants in the *APOE* gene can alter a person's lifetime risk for the disease by a factor of 3–4, for each copy of the APOE4 genotype carried. Around 20 other genetic loci, discovered from genome-wide association studies comparing Alzheimer's patients to controls, consistently affect AD risk by around 2%–20%. There is great interest in understanding which brain systems and networks these genes affect, and at what times during the lifespan, and if these effects can be influenced by interventions.

A vast array of candidate SNPs in genes of interest have been studied that may be hypothesized to affect brain connectivity and microstructure—from growth factors and neurotrophins, to genes involved in axon guidance and cell proliferation, to genes implicated in apoptosis. Around 10 years ago, the most common type of imaging genetics study would typically assess common variants in one or a few candidate genes, and assess their effects on brain metrics derived from MRI, diffusion MRI, or functional MRI, typically in under a thousand people (Chiang et al., 2011; Braskie et al., 2011, 2012; Jahanshad et al., 2012; Dennis et al., 2011; Scott-Van Zeeland et al., 2010; Filippini et al., 2009; Rudie et al., 2012; Brown et al., 2011; Canuet et al., 2012; Liu et al., 2010; Tunbridge et al., 2013; Trachtenberg et al., 2012).

As became clear in attempts to replicate these studies, the associations between specific candidate SNPs and brain MRI measures were often shown to not replicate when tested in independent samples, suggesting that many results might be considered spurious or not generalize beyond the cohort being studied (Jahanshad et al., 2017). More likely, the large number of candidate genes and brain measures available led to "alpha inflation" or "winner's curse" where an initially successful study did not replicate in future independent data. The psychiatric genetics field underwent a similar "replication crisis" in regard to candidate genes that were thought for some time to associate with schizophrenia; the field of imaging genetics and psychiatric genetics both evolved to larger scale, more rigorous and better powered studied that searched the whole genome or all common variants in an unbiased way. Due to the multiple comparisons corrections inherent in searching over a million genetic loci for associations with brain metrics, the field set a significant level of 0.05 divided by the effective number of statistical tests; as a result, most well powered studies reporting a genetic association with a brain measure—with the possible exception of APOE—tended to analyze over 1000 subjects, and sometimes 20,000–40,000 (Hibar et al., 2015; Satizabal et al., 2017).

GENOME-WIDE SEARCHES AND BOOSTING STATISTICAL POWER TO ADDRESS SMALL EFFECT SIZES

In under 1000 people, Smith and Nichols (Smith and Nichols, 2018) have argued that there is very limited power to detect genetic variants (or any other factors) that account for under 1% of the variance in any brain measure, and most SNP effects are known to fall in this range, even in cases where they collectively account for over a third of the variance in a trait. Medland et al. (2014) noted that large imaging consortia, such as ENIGMA and CHARGE, were beginning to discover common genetic variants that affected subcortical brain volumes at significant levels lower than $P = 10^{-12}$ (Hibar et al., 2015; Satizabal et al., 2017; Adams et al., 2016) meaning that it was feasible to search thousands of traits for genetic influences at over a million genetic loci, while applying appropriate safeguards against false positive findings. By performing GWAS across a population at each voxel in a set of brain images (Fig. 1), or at each connection in the connectome (Jahanshad et al., 2013a,b), the concept of *connectome-wide*

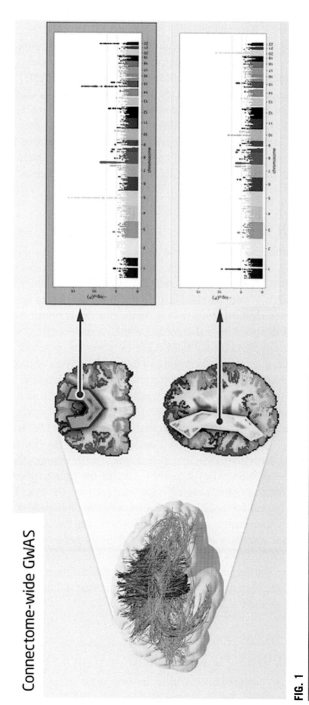

FIG. 1

A genome-wide association screen can be conducted at for global network properties, nodal measures, or every element of a connectivity matrix, as depicted in the figure, leading to a connectome-wide GWAS.

Modified from Medland, S.E., Jahanshad, N., Neale, B.M., Thompson, P.M., 2014. Whole-genome analyses of whole-brain data: working within an expanded search space. Nat. Neurosci. 17 (6), 791–800.

genome-wide search allowed a well-powered and unbiased search for genetic variants that affect brain connectivity. The mathematics of dimensionality reduction that can make this kind of search more efficient is beyond the scope of this chapter, and includes many methods from classical multivariate statistics, information theory, as and machine learning; an interested reader can consult (Thompson et al., 2013).

Recent genome-wide searches across an array of brain-based measures have identified many individual variants associated with variation in brain structure. However, a lack of significant genetic variant associations with brain functional connectivity have been surprising (Elliott et al., 2018), suggesting perhaps that function itself as approximated by BOLD fMRI may have high *polygenicity*, having a higher number of variants explaining population differences, each with smaller effect-sizes.

GENE EXPRESSION NETWORKS

The genetic analysis of brain microstructure and brain networks may also involve the examination of other kinds of genetic and epigenetic variation, as well as other molecular markers that assess proteins and metabolites relevant to brain health and brain aging. These may include measures of beta-amyloid or tau protein levels, plasma lipids, cholesterol and other markers of cardiovascular disease, endocrine markers such as thyroid hormones or cortisol, or inflammatory and immune system markers such as cytokines. A person's genetic code is largely stable throughout life, but there are epigenetic modifications throughout life that affect gene expression and disease risk. Some of this disease risk may be mediated by brain metrics that can be measured from brain scans.

In an effort to relate variations in brain connectivity to some of these more dynamic markers of molecular markers and gene expression, Fulcher and Fornito (2016) related the neuronal connectivity of 213 brain regions to a measure of the "transcriptional coupling" across 17,642 genes, between each pair of regions (Fulcher and Fornito, 2016). They also concentrated on a small number of highly connected neural elements that act as network hubs. They hypothesized that the transcriptional coupling would be highest for pairs of connected hubs, intermediate for links between hubs and nonhubs, and lowest for connected pairs of nonhubs; this pattern was, in fact, borne out by their experimental data. They theorized that the high transcriptional coupling associated with hub connectivity is driven by genes regulating the oxidative synthesis and metabolism of ATP—the major source of energy supporting neuronal communication. They argued that brain hubs display a tight coordination of gene expression, often over long anatomical distances, that may relate to the metabolic requirements of connected network elements.

TRANSLATING FINDINGS TO THE CLINIC

The genetic influences on connectivity metrics are important to understand from a basic neuroscience standpoint, but they may also be informative for understanding the causes of differences in patient populations, including patient outcomes, responses to specific treatments, and trajectories of recovery. In fact, the whole field of pharmacogenomics is motivated by the premise that a person's treatment response may depend in part on individual genetic variation, so it is important to model and understand the effects of genetic variation on clinical measures of disease burden, such as connectivity, or disturbances in the brain's networks.

Even in cases where a disease has a complex etiology—such as infectious disease, MDD, or PTSD—an individual's genetic profile may still influence their disease severity, treatment response, and prognosis. Below we give an example of how genetic analysis of brain connectivity has been used to understand clinical outcomes in traumatic brain injury.

ALTERED CONNECTIVITY IN TRAUMATIC BRAIN INJURY

Traumatic brain injury (TBI) is an environmental factor that can interact with genetics to affect brain connectivity. Both structural and functional connectivity can be affected postinjury, contributing to the physical, emotional, and cognitive disruptions that some patients experience. There is tremendous heterogeneity in outcome postinjury, only partially explained by injury variables. Preexisting conditions, differences in treatment, family conditions, and genetics all likely play a role in this heterogeneity.

TBI leads to widespread network disruptions, both structural and functional (Sharp et al., 2014; Hayes et al., 2016). Recently there has been a wealth of research using dMRI to identify and track structural disruptions following TBI, due in large part to its increased sensitivity to traumatic axonal injury (TAI) compared to conventional imaging methods like computed tomography (CT) (Hulkower et al., 2013). Poorer WM organization has been shown following moderate-severe TBI even in the absence of extensive cortical damage (Spitz et al., 2013; Kinnunen et al., 2010; Bendlin et al., 2008), and may be present in mild TBI as well, but those results have been mixed (Spitz et al., 2013; Aoki et al., 2012; Yuh et al., 2014; Shenton et al., 2012). Blast-related TBI, a signature injury of the recent wars in Iraq and Afghanistan, most often qualifies as "mild," but is associated with disrupted WM organization (Wilde et al., 2015; Mac Donald et al., 2011). Increased attention has been given recently as well to the impact of repetitive mild TBI, which is similarly associated with disrupted WM organization (Koerte et al., 2015). Disruptions to functional connectivity have also been reported postinjury. The small-world organization of the network is disturbed (Pandit et al., 2013), but also shows evidence of recovery longitudinally (Nakamura et al., 2009). (Gratton et al., 2012) showed that focal lesions disrupt the global modular organization of the functional connectome. Similar disruptions to modular structure have been shown following blast-related injury (Han et al., 2014), and may be greater with closer proximity to the blast (Robinson et al., 2015). Repetitive head injuries, even at the subconcussive level, can also lead to alterations in functional connectivity (Abbas et al., 2015).

In pediatric patients, structural connectivity is disrupted, especially in the corpus callosum and long anterior-posterior tracts. These disruptions can be seen as early as 2 months postinjury, and can persist for years postinjury (Dennis et al., 2018). There is considerable heterogeneity in outcome postinjury in children, however, which is not fully explained by injury variables. One longitudinal study found an effect of age on outcome (Ewing-Cobbs et al., 2016), while another identified subgroups of patients with diverging outcomes based on an EEG measure of callosal integrity (Dennis et al., 2017). The existing data suggest that structural connectivity deficits are driven by myelin pathology, more so than axonal pathology (Dennis et al., 2015; Wilde et al., 2012). Functional connectivity is also disrupted, with decreased connectivity among emotion regulation structures, the DMN, and motor networks (Stephens et al., 2017), and increased activity between attention networks (Newsome et al., 2013; Risen et al., 2015; Diez et al., 2017). These studies show both functional disruption including evidence of increased attention necessary to complete tasks and possible compensatory mechanisms.

THE INTERACTION BETWEEN GENETICS AND TBI

While TBI is an environmentally acquired variable, we are increasingly understanding the role that genetics plays in the injury and recovery process. The majority of studies that have examined the impact of genetics on outcome following TBI have used a candidate gene approach, focusing on polymorphisms that prior studies have linked with cognitive function and emotion, such as *APOE*, *COMT*, *BDNF*, and dopamine system genes, along with interleukin genes that play a role in the postinjury inflammatory response (Dardiotis et al., 2010; Jordan, 2007). Likely due to the very large sample size necessary, there have been no GWAS studies of outcome after TBI (Lipsky and Lin, 2015; Kurowski et al., 2012).

Most of the research on the impact of genetics on outcome after TBI has focused on *APOE*, in part because of the large effect size associated with the ε4allele, but also because TBI patients are at an increased risk of developing dementia later on (Moretti et al., 2012). The DoD ADNI study of Vietnam veterans is a multisite effort in the United States to understand how TBI and PTSD confer risk for dementia (Weiner et al., 2014). While a number of studies have reported an association between *APOE* genotype and outcome (Jordan et al., 1997; Moran et al., 2009), others have found no relationship (Hiekkanen et al., 2009). One study found that variations in the *BDNF* gene were association with memory function after injury (McAllister et al., 2012). One metaanalysis found that *APOE* genotype was associated with 6-month outcome (Zhou et al., 2008), while another recent review of the impact of *APOE* on recovery found that the ε4allele was most often reported as detrimental in severe TBI studies, while those covering mild and moderate injuries less often reported a significant effect (Lawrence et al., 2015). One large study found an interaction between *APOE* genotype, age, and outcome, with unfavorable outcome most pronounced in *APOE* ε4carriers under 15 years old (Wood, 2017). The issue of sensitive periods for brain injury has not been resolved in the literature thus far. A key factor to consider in future research on this topic will be the contribution of genetics, as well. Variation in a SNP associated with the *DRD2* gene has been associated with episodic memory performance after injury (McAllister et al., 2005), while variation in the *COMT* gene is linked with executive function after injury (Lipsky et al., 2005).

There have been a number of studies showing the influence of genetics on outcome postinjury; however, very few have also included neuroimaging. (Isoniemi et al., 2006) examined the impact of *APOE* genotype on the volume of the hippocampus and lateral ventricles decades postinjury, reporting null results. Similarly, (Hiekkanen et al., 2007) reported no significant contribution of *APOE* genotype to MRI measures 1 year postinjury. With sample sizes between 30 and 60, however, these were likely underpowered analyses. There have been no published studies examining the genetics of structural or functional connectivity postinjury.

While there have been a number of studies showing altered structural and functional connectivity after TBI (Dennis et al., 2018; Hulkower et al., 2013; Hayes et al., 2016; Wilde et al., 2015), as well as others examining the link between genetics and recovery postinjury, none have looked at the intersection of genetics and connectomics postinjury. This is an important avenue for future research. The lack of research in this area is likely due in large part to the large sample sizes that are necessary for well-powered genetic analysis. Several large multisite studies have great promise to address this gap: TRACK-TBI (Transforming Research and Clinical Knowledge in Traumatic Brain Injury; (Yue et al., 2013)) and CENTER-TBI (Collaborative European NeuroTrauma Effectiveness Research in Traumatic Brain Injury; (Maas et al., 2015)), along with the Brain Injury working group of ENIGMA (Enhancing NeuroImaging Genetics through Metaanalysis). In the pediatric TBI population, there has been even less work including genetics. Pediatric TBI in general has received less attention than adult TBI, and researchers may have been less likely to request blood samples in this population. As in adults, however, genetics holds great promise to explain some of the variability in outcome. Developing populations are vulnerable to long-term disability following injury (>61% experience disability after moderate or severe TBI [Kirch, 2018]), as maturational processes can be interrupted. This also means that effective intervention can make a profound difference; interventions that help a child get back on track will be magnified over time, as developmental targets are met and risk of disability decreases.

FUTURE DIRECTIONS

The future of imaging genetics is at an important crossroads where there is now enough data from human populations worldwide to enable well-powered replicated studies of genetic influences on brain connectivity. The setbacks and misfires of many earlier candidate gene studies tended to fill the literature with findings that were hard to verify or did not stand up to testing in independent samples; the current size of biobanks and global imaging genetics efforts such as ENIGMA and CHARGE enable multiple traits to be studied at once without the risk of large proportions of unreplicable false discoveries.

A second key direction is the move from metrics of genetic variation at the SNP level to more holistic models of multiple gene effects—or gene networks—working on multiple brain systems.

There is a natural alliance between the field of network science and both connectomics and genetics in discovering these patterns with a realistic level of complexity. Indeed, much of the graph theoretic development that has energized the connectomics field has found application in understanding interactions between genes, and future developments in network science are sure to advance both fields.

Much of the molecular variation that drives functional connectivity may operate in a dynamic fashion that is not yet seen in large-scale population studies; with enough data—and with targeted experimental designs—it should soon be possible to synergize efforts in proteomics and biomarker discovery so that the whole armory of experimental molecular methods can used to probe the brain's networks.

ACKNOWLEDGMENTS

The authors acknowledge NIH grant funding from the "Big Data to Knowledge" Program (U54 EB020403), and a NINDS K award (to E.L.D., K99NS096116).

REFERENCES

Abbas, K., Shenk, T.E., Poole, V.N., Breedlove, E.L., Leverenz, L.J., Nauman, E.A., Talavage, T.M., Robinson, M.E., 2015. Alteration of default mode network in high school football athletes due to repetitive subconcussive mild traumatic brain injury: a resting-state functional magnetic resonance imaging study. Brain Connect. 5 (2), 91–101.

Adams, H.H., Hibar, D.P., Chouraki, V., Stein, J.L., Nyquist, P.A., Rentería, M.E., Trompet, S., et al., 2016. Novel genetic loci underlying human intracranial volume identified through genome-wide association. Nat. Neurosci. 19 (12), 1569–1582. nature.com.

Adhikari, B.M., Jahanshad, N., Shukla, D., Glahn, D.C., Blangero, J., Reynolds, R.C., Cox, R.W., et al., 2018. In: Heritability estimates on resting state fMRI data using ENIGMA analysis pipeline. Pacific Symposium on Biocomputing, vol. 23, pp. 307–318.

Anderson, J.S., Nielsen, J.A., Ferguson, M.A., Burback, M.C., Cox, E.T., Dai, L., Gerig, G., Edgin, J.O., Korenberg, J.R., 2013. Abnormal brain synchrony in down syndrome. NeuroImage Clin. 2, 703–715.

Aoki, Y., Inokuchi, R., Gunshin, M., Yahagi, N., Suwa, H., 2012. Diffusion tensor imaging studies of mild traumatic brain injury: a meta-analysis. J. Neurol. Neurosurg. Psychiatry 83 (9), 870–876.

Arlinghaus, L.R., Thornton-Wells, T.A., Dykens, E.M., Anderson, A.W., 2011. Alterations in diffusion properties of white matter in Williams syndrome. Magn. Reson. Imaging 29, 1165–1174.

Avery, S.N., Thornton-Wells, T.A., Anderson, A.W., Blackford, J.U., 2012. White matter integrity deficits in prefrontal-amygdala pathways in williams syndrome. Neuroimage 59 (2), 887–894. Elsevier.

Barnea-Goraly, N., Eliez, S., Hedeus, M., Menon, V., White, C.D., Moseley, M., Reiss, A.L., 2003a. White matter tract alterations in fragile X syndrome: preliminary evidence from diffusion tensor imaging. Am. J. Med. Genet. B. Neuropsychiatr. Genet. 118B, 81–88.

Barnea-Goraly, N., Menon, V., Krasnow, B., Ko, A., Reiss, A., Eliez, S., 2003b. Investigation of white matter structure in Velocardiofacial syndrome: a diffusion tensor imaging study. Am. J. Psychiatry 160, 1863–1869.

Bassett, A.S., Chow, E.W.C., 1999. 22q11 deletion syndrome: a genetic subtype of schizo-phrenia. Biol. Psychiatry 46 (7), 882–891. Elsevier.

Bellugi, U., Lichtenberger, L., Jones, W., Lai, Z., George, M.S., 2001. The neurocognitive profile of Williams syndrome: a complex pattern of strengths and weaknesses. J. Cogn. Neurosci 12, 7–29.

Bendlin, B.B., Ries, M.L., Lazar, M., Alexander, A.L., Dempsey, R.J., Rowley, H.A., Sherman, J.E., Johnson, S.C., 2008. Longitudinal changes in patients with traumatic brain injury as-sessed with diffusion-tensor and volumetric imaging. Neuroimage 42 (2), 503–514.

Bohanna, I., Georgiou-Karistianis, N., Egan, G.F., 2011a. Connectivity-based segmentation of the striatum in Huntington's disease: vulnerability of motor pathways. Neurobiol. Dis. 42 (3), 475–481.

Bohanna, I., Georgiou-Karistianis, N., Sritharan, A., Asadi, H., Johnston, L., Churchyard, A., Egan, G., 2011b. Diffusion tensor imaging in Huntington's disease reveals distinct patterns of white matter degeneration associated with motor and cognitive deficits. Brain Imaging Behav. 5 (3), 171–180.

Braskie, M.N., Jahanshad, N., Stein, J.L., Barysheva, M., McMahon, K.L., De Zubicaray, G.I., Martin, N.G., et al., 2011. Common Alzheimer's disease risk variant within the CLU gene affects white matter microstructure in young adults. J. Neurosci. 31 (18), 6764–6770.

Braskie, M.N., Jahanshad, N., Stein, J.L., Barysheva, M., Johnson, K., Mcmahon, K.L., de Zubicaray, G.I., et al., 2012. Relationship of a variant in the NTRK1 gene to white matter mi-crostructure in young adults. J. Neurosci. http://www.jneurosci.org/content/32/17/5964.short.

Bray, S., Dunkin, B., Hong, D.S., Reiss, A.L., 2011. Reduced functional connectivity during working memory in turner syndrome. Cereb. Cortex 21, 2471–2481.

Bray, S., Hoeft, F., Hong, D.S., Reiss, A.L., 2012. Aberrant functional network recruitment of posterior parietal cortex in turner syndrome. Hum. Brain Mapp. https://doi.org/10.1002/hbm.22131.

Brown, J.A., Terashima, K.H., Burggren, A.C., Ercoli, L.M., Miller, K.J., Small, G.W., Bookheimer, S.Y., 2011. Brain network local interconnectivity loss in aging APOE-4 allele carriers. Proc. Natl. Acad. Sci. 108 (51), 20760–20765.

Bruno, J.L., Hadi Hosseini, S.M., Saggar, M., Quintin, E.-M., Raman, M.M., Reiss, A.L., 2017. Altered brain network segregation in fragile X syndrome revealed by structural con-nectomics. Cereb. Cortex 27 (3), 2249–2259. academic.oup.com.

Budisavljevic, S., Kawadler, J.M., Dell'Acqua, F., Rijsdijk, F.V., Kane, F., Picchioni, M., McGuire, P., et al., 2016. Heritability of the limbic networks. Soc. Cogn. Affect. Neurosci. 11 (5), 746–757.

Canuet, L., Tellado, I., Couceiro, V., Fraile, C., Fernandez-Novoa, L., Ishii, R., Takeda, M., Cacabelos, R., 2012. Resting-state network disruption and APOE genotype in Alzheimer's disease: a lagged functional connectivity study. PLoS One 7 (9), e46289.

Cassidy, S.B., 1997. Prader-Willi Syndrome. J. Med. Genet. 34 (11), 917–923. jmg.bmj.com.

Catani, M., de Schotten, M.T., 2012. Atlas of Human Brain Connections. Oxford University Press, Oxford.

Chiang, M.C., Barysheva, M., Shattuck, D.W., Lee, A.D., Madsen, S.K., Avedissian, C., Klunder, A.D., et al., 2009. Genetics of brain fiber architecture and intellectual perfor-mance. J. Neurosci. 29 (7), 2212–2224.

Chiang, M.-C., Mcmahon, K.L., de Zubicaray, G.I., Martin, N.G., Hickie, I., Toga, A.W., Wright, M.J., Thompson, P.M., 2011. Genetics of white matter development: a DTI study of 705 twins and their siblings aged 12 to 29. Neuroimage 54 (3), 2308–2317.

Couvy-Duchesne, B., Blokland, G.A.M., Hickie, I.B., Thompson, P.M., Martin, N.G., de Zubicaray, G.I., McMahon, K.L., Wright, M.J., 2014. Heritability of head motion during resting state functional MRI in 462 healthy twins. Neuroimage 102 (Pt 2), 424–434.

Dardiotis, E., Fountas, K.N., Dardioti, M., Xiromerisiou, G., Kapsalaki, E., Tasiou, A., Hadjigeorgiou, G.M., 2010. Genetic association studies in patients with traumatic brain injury. Neurosurg. Focus 28 (1), E9.

Debbané, M., Lazouret, M., Lagioia, A., Schneider, M., Van De Ville, D., Eliez, S., 2012. Resting-state networks in adolescents with 22q11.2 deletion syndrome: associations with prodromal symptoms and executive functions. Schizophr. Res. 139 (1–3), 33–39. schresjournal.com.

Demuru, M., Gouw, A.A., Hillebrand, A., Stam, C.J., van Dijk, B.W., Scheltens, P., Tijms, B.M., et al., 2017. Functional and effective whole brain connectivity using magnetoencephalography to identify monozygotic twin pairs. Sci. Rep. 7 (1), 9685.

Dennis, E.L., Jahanshad, N., Rudie, J.D., Brown, J.A., Johnson, K., Mcmahon, K.L., de Zubicaray, G.I., et al., 2011. Altered structural brain connectivity in healthy carriers of the autism risk gene, CNTNAP2. Brain Connect. 1 (6), 447–459.

Dennis, E.L., Jin, Y., Villalon-Reina, J., Zhan, L., Kernan, C., Babikian, T., Mink, R., Babbitt, C., Johnson, J., Giza, C.C., 2015. White matter disruption in moderate/severe pediatric traumatic brain injury: advanced tract-based analyses. NeuroImage Clin. 7, 493–505.

Dennis, E.L., Rashid, F., Ellis, M.U., Babikian, T., Villalon-Reina, J.E., Jin, Y., Olsen, et al., 2017. Diverging white matter trajectories in children after traumatic brain injury: the RAPBI study. Neurology 88 (15), 1392–1399.

Dennis, E.L., Babikian, T., Giza, C.C., Thompson, P.M., Asarnow, R.F., 2018. Neuroimaging of the injured pediatric brain: methods and new lessons. Neuroscientist. [Epub ahead of print]. https://doi.org/10.1177/1073858418759489.

Di Paola, M., Luders, E., Cherubini, A., Sanchez-Castaneda, C., Thompson, P.M., Toga, A.W., Caltagirone, C., et al., 2012. Multimodal MRI analysis of the Corpus callosum reveals white matter differences in Presymptomatic and early Huntington's disease. Cereb. Cortex 22 (12), 2858–2866.

Diez, I., Drijkoningen, D., Stramaglia, S., Bonifazi, P., Marinazzo, D., Gooijers, J., Swinnen, S.P., Cortes, J.M., 2017. Enhanced prefrontal functional–structural networks to support postural control deficits after traumatic brain injury in a pediatric population. Netw. Neurosci. 1 (2), 116–142.

Domínguez, D., Juan, F., Egan, G.F., Gray, M.A., Poudel, G.R., Churchyard, A., Chua, P., Stout, J.C., Georgiou-Karistianis, N., 2013. Multi-modal neuroimaging in Premanifest and early Huntington's disease: 18 month longitudinal data from the IMAGE-HD study. PLoS One 8 (9), e74131.

Dumas, E.M., van den Bogaard, S.J.A., Ruber, M.E., Reilman, R.R., Stout, J.C., Craufurd, D., Hicks, S.L., et al., 2012. Early changes in white matter pathways of the sensorimotor cortex in Premanifest Huntington's disease. Hum. Brain Mapp. 33 (1), 203–212.

Dumas, E.M., van den Bogaard, S.J.A., Hart, E.P., Soeter, R.P., van Buchem, M.A., van der Grond, J., Rombouts, S.A.R.B., Roos, R.A.C., TRACK-HD investigator group, 2013. Reduced functional brain connectivity prior to and after disease onset in huntington's disease. NeuroImage Clin. 2, 377–384. Elsevier.

Elliott, L., Sharp, K., Alfaro-Almagro, F., Shi, S., Miller, K., Douaud, G., Marchini, J., Smith, S., 2018. Genome-wide association studies of brain structure and function in the UK Biobank. Cold Spring Harbor Laboratory. https://doi.org/10.1101/178806.

Espinoza, F.A., Turner, J.A., Vergara, V.M., Miller, R.L., Mennigen, E., Liu, J., Misiura, M.B., et al., 2018. Whole-brain connectivity in a large study of huntington's disease gene mutation carriers and healthy controls. Brain Connect. online.liebertpub.com. https://doi.org/10.1089/brain.2017.0538.

Ewing-Cobbs, L., Johnson, C.P., Juranek, J., DeMaster, D., Prasad, M., Duque, G., Kramer, L., Cox, C.S., Swank, P.R., 2016. Longitudinal diffusion tensor imaging after pediatric traumatic brain injury: Impact of age at injury and time since injury on pathway integrity. Hum. Brain Mapp. 37 (11), 3929–3945.

Fenoll, R., Pujol, J., Esteba-Castillo, S., de Sola, S., Ribas-Vidal, N., García-Alba, J., Sánchez-Benavides, G., et al., 2017. Anomalous white matter structure and the effect of age in down syndrome patients. J. Alzheim. Dis. 57 (1), 61–70. content.iospress.com.

Filippini, N., Rao, A., Wetten, S., Gibson, R.A., Borrie, M., Guzman, D., Kertesz, A., et al., 2009. Anatomically-distinct genetic associations of APOE epsilon4 allele load with regional cortical atrophy in Alzheimer's disease. Neuroimage 44 (3), 724–728.

Fornito, A., Zalesky, A., Bassett, D.S., Meunier, D., Ellison-Wright, I., Yucel, M., Wood, S.J., et al., 2011. Genetic influences on cost-efficient organization of human cortical functional networks. J. Neurosci. 31 (9), 3261–3270.

Fulcher, B.D., Fornito, A., 2016. A transcriptional signature of hub connectivity in the mouse connectome. Proc. Natl. Acad. Sci. U. S. A. 113 (5), 1435–1440.

Gargouri, F., Messé, A., Perlbarg, V., Valabregue, R., McColgan, P., Yahia-Cherif, L., Fernandez-Vidal, S., et al., 2016. Longitudinal changes in functional connectivity of cortico-basal ganglia networks in manifests and premanifest huntington's disease. Hum. Brain Mapp. 37 (11), 4112–4128. Wiley Online Library.

Geng, X., Prom-Wormley, E.C., Perez, J., Kubarych, T., Styner, M., Lin, W., Neale, M.C., Gilmore, J.H., 2012. White matter heritability using diffusion tensor imaging in neonatal brains. Twin Res. Hum. Genet. 15 (3), 336–350.

Glahn, D.C., Winkler, A.M., Kochunov, P., Almasy, L., Duggirala, R., Carless, M.A., Curran, J.C., et al., 2010. Genetic control over the resting brain. Proc. Natl. Acad. Sci. U. S. A. 107 (3), 1223–1228.

Gratton, C., Nomura, E.M., Pérez, F., D'Esposito, M., 2012. Focal brain lesions to critical locations cause widespread disruption of the modular organization of the brain. J. Cogn. Neurosci. 24 (6), 1275–1285.

Gregory, S., Cole, J.H., Farmer, R.E., Rees, E.M., Roos, R.A.C., Sprengelmeyer, R., Durr, A., et al., 2015. Longitudinal diffusion tensor imaging shows progressive changes in white matter in Huntington's disease. J. Huntington's Dis. 4 (4), 333–346.

Guadalupe, T., Mathias, S.R., vanErp, T.G.M., Whelan, C.D., Zwiers, M.P., et al., 2017. Human subcortical brain asymmetries in 15,847 people worldwide reveal effects of age and sex. Brain Imaging Behav. 11 (5), 1497–1514.

Gunbey, H.P., Bilgici, M.C., Aslan, K., Has, A.C., Ogur, M.G., Alhan, A., Incesu, L., 2017. Structural Brain Alterations of Down's Syndrome in Early Childhood Evaluation by DTI and Volumetric Analyses. Eur. Radiol. 27 (7), 3013–3021. Springer.

Haas, B.W., Barnea-Goraly, N., Lightbody, A., Patnaik, S.S., Hoeft, F., Hazlett, H., Piven, J., Reiss, A.L., 2009. Early white-matter abnormalities of the ventral Frontostriatal pathway in fragile X syndrome. Dev. Med. Child Neurol. 51, 593–599.

Haas, B.W., Barnea-Goraly, N., Sheau, K.E., Yamagata, B., Ullas, S., Reiss, A.L., 2014. Altered microstructure within social-cognitive brain networks during childhood in Williams syndrome. Cereb. Cortex 24 (10), 2796–2806. academic.oup.com.

Hagerman, R.J., Berry-Kravis, E., Hazlett, H.C., Jr, D.B.B., Moine, H., Frank Kooy, R., Tassone, F., et al., 2017. Fragile X Syndrome. Nat. Rev. 3, 17065. nature.com.

Hall, S.S., Jiang, H., Reiss, A.L., Greicius, M.D., 2013. Identifying large-scale brain networks in fragile X syndrome. JAMA Psychiat. 70 (11), 1215–1223. jamanetwork.com.

Hall, S.S., Dougherty, R.F., Reiss, A.L., 2016. Profiles of aberrant white matter microstructure in fragile X syndrome. NeuroImage Clin. 11, 133–138. Elsevier.

Han, K., Mac Donald, C.L., Johnson, A.M., Barnes, Y., Wierzechowski, L., Zonies, D., John, O., et al., 2014. Disrupted modular organization of resting-state cortical functional connectivity in U.S. military personnel following concussive 'mild' blast-related traumatic brain injury. Neuroimage 84, 76–96.

Harrington, D.L., Rubinov, M., Durgerian, S., Mourany, L., Reece, C., Koenig, K., Ed, B., et al., 2015. Network topology and functional connectivity disturbances precede the onset of Huntington's disease. Brain J. Neurol. 138 (Pt 8), 2332–2346. academic.oup.com.

Harrington, D.L., Long, J.D., Durgerian, S., Mourany, L., Koenig, K., Bonner-Jackson, A., Paulsen, J.S., PREDICT-HD Investigators of the Huntington Study Group, S.M., Rao, 2016. Cross-sectional and longitudinal multimodal structural imaging in prodromal huntington's disease. Movem. Disord. 31 (11), 1664–1675.

Hayes, J.P., Bigler, E.D., Verfaellie, M., 2016. Traumatic brain injury as a disorder of brain connectivity. J. Int. Neuropsychol. Soc. 22 (2), 120–137.

Hibar, D.P., Stein, J.L., Renteria, M.E., Arias-Vasquez, A., Desrivieres, S., Jahanshad, N., Toro, R., et al., 2015. Common genetic variants influence human subcortical brain structures. Nature 520 (7546), 224–229.

Hibar, D.P., Adams, H.H.H., Jahanshad, N., Chauhan, G., Stein, J.L., Hofer, E., Renteria, M.E., et al., 2017. Novel genetic loci associated with hippocampal volume. Nat. Commun. 8, 13624. nature.com.

Hiekkanen, H., Kurki, T., Brandstack, N., Kairisto, V., Tenovuo, O., 2007. MRI changes and ApoE genotype, a prospective 1-year follow-up of traumatic brain injury: a pilot study. Brain Inj. 21 (12), 1307–1314. Taylor & Francis.

Hiekkanen, H., Kurki, T., Brandstack, N., Kairisto, V., Tenovuo, O., 2009. Association of injury severity, MRI-results and ApoE genotype with 1-year outcome in mainly mild TBI: a preliminary study. Brain Inj. 23 (5), 396–402.

Hoeft, F., Barnea-Goraly, N., Haas, B.W., Golarai, G., Ng, D., Mills, D., Korenberg, J., Bellugi, U., Galaburda, A., Reiss, A.L., 2007. More is not always better: Increased fractional anisotropy of superior longitudinal fasciculus associated with poor visuospatial abilities in Williams syndrome. J. Neurosci. 27 (44), 11960–11965.

Holzapfel, M., Barnea-Goraly, N., Eckert, M.A., Kesler, S.R., Reiss, A.L., 2006. Selective alterations of white matter associated with visuospatial and sensorimotor dysfunction in turner syndrome. J. Neurosci. 26 (26), 7007–7013.

Hulkower, M.B., Poliak, D.B., Rosenbaum, S.B., Zimmerman, M.E., Lipton, M.L., 2013. A decade of DTI in traumatic brain injury: 10 years and 100 articles later. Am. J. Neuroradiol. https://doi.org/10.3174/ajnr.A3395.

Isoniemi, H., Kurki, T., Tenovuo, O., Kairisto, V., Portin, R., 2006. Hippocampal volume, brain atrophy, and APOE genotype after traumatic brain injury. Neurology 67 (5), 756–760.

Jabbi, M., Shane Kippenhan, J., Kohn, P., Marenco, S., Mervis, C.B., Morris, C.A., Meyer-Lindenberg, A., Berman, K.F., 2012. The Williams syndrome chromosome 7q11.23hemideletion confers hypersocial, anxious personalitycoupled with altered insula structure and function. Proc. Natl. Acad. Sci. E860–E866.

Jahanshad, N., Lee, A., Barysheva, M., McMahon, K., 2010. Genetic influences on brain asymmetry: a DTI study of 374 twins and siblings. Neuroimage. http://www.com/science/article/pii/S1053811910006488.

Jahanshad, N., Kohannim, O., Hibar, D.P., Stein, J.L., Mcmahon, K.L., de Zubicaray, G.I., Medland, S.E., et al., 2012. Brain structure in healthy adults is related to serum transferrin and the H63D polymorphism in the HFE gene. Proc. Natl. Acad. Sci. 109 (14), E851–E859.

Jahanshad, N., Kochunov, P.V., Sprooten, E., Mandl, R.C., Nichols, T.E., Almasy, L., Blangero, J., et al., 2013a. Multi-site genetic analysis of diffusion images and voxelwise heritability analysis: a pilot project of the ENIGMA–DTI working group. Neuroimage 81, 455–469.

Jahanshad, N., Rajagopalan, P., Hua, X., Hibar, D.P., Nir, T.M., Toga, A.W., Jack, C.R., et al., 2013b. Genome-wide scan of healthy human connectome discovers spon1 gene variant influencing dementia severity. Proc. Natl. Acad. Sci. http://www.pnas.org/content/early/2013/02/28/1216206110.short.

Jahanshad, N., Ganjgahi, H., Bralten, J., den Braber, A., Faskowitz, J., Knodt, A., et al., 2017. Do Candidate Genes Affect the Brain's White Matter Microstructure? Large-Scale Evaluation of 6,165 Diffusion MRI Scans. Cold Spring Harbor Laboratory.

Jalbrzikowski, M., Villalon-Reina, J.E., Karlsgodt, K.H., Senturk, D., Chow, C., Thompson, P.M., Bearden, C.E., 2014. Altered white matter microstructure is associated with social cognition and psychotic symptoms in 22q11.2 microdeletion syndrome. Front. Behav. Neurosci. 8, 393. frontiersin.org.

Jordan, A.D., Relkin, N.R., Ravdin, L.D., Jacobs, A.R., 1997. Apolipoprotein E e4 associated with chronic traumatic brain injury in boxing. JAMA 278 (2), 136–140.

Jordan, B.D., 2007. Genetic influences on outcome following traumatic brain injury. Neurochem. Res. 32 (4-5), 905–915.

Kates, W.R., Olszewski, A.K., Gnirke, M.H., Kikinis, Z., Nelson, J., Antshel, K.M., Fremont, W., et al., 2015. White matter microstructural abnormalities of the cingulum bundle in youths with 22q11.2 deletion syndrome: Associations with medication, neuropsychological function, and prodromal symptoms of psychosis. Schizophr. Res. 161 (1), 76–84. schres-journal.com.

Kinnunen, K.M., Greenwood, R., Powell, J.H., Leech, R., Hawkins, P.C., Bonnelle, V., Patel, M.C., Counsell, S.J., Sharp, D.J., 2010. White Matter Damage and Cognitive Impairment after Traumatic Brain Injury. Brain J. Neurol. . awq347.

Kirch, W. (Ed.), 2018. The management of traumatic brain injury in children: opportunities for action. Encyclopedia of Public HealthSpringer Netherlands, Dordrecht.

Kochunov, P., Glahn, D.C., Lancaster, J.L., Winkler, A.M., Smith, S., Thompson, P.M., Almasy, L., Duggirala, R., Fox, P.T., Blangero, J., 2010. Genetics of microstructure of cerebral white matter using diffusion tensor imaging. Neuroimage 53 (3), 1109–1116.

Koenig, K.A., Lowe, M.J., Harrington, D.L., Lin, J., Durgerian, S., Mourany, L., Paulsen, J.S., Rao, S.M., PREDICT-HD Investigators of the Huntington Study Group, 2014. Functional connectivity of primary motor cortex is dependent on genetic burden in prodromal huntington disease. Brain Connect. 4 (7), 535–546. online.liebertpub.com.

Koerte, I.K., Lin, A.P., Willems, A., Muehlmann, M., Hufschmidt, J., Coleman, M.J., Green, I., et al., 2015. A review of neuroimaging findings in repetitive brain trauma. Brain Pathol. 25 (3), 318–349.

Kurowski, B., Martin, L.J., Wade, S.L., 2012. Genetics and outcomes after traumatic brain injury (TBI): what do we know about pediatric TBI? J. Pediatr. Rehabil. Med. 5 (3), 217–231.

Lawrence, D.W., Comper, P., Hutchison, M.G., Sharma, B., 2015. The role of apolipoprotein E Episilon (ε)-4 allele on outcome following traumatic brain injury: a systematic review. Brain Inj. 29 (9), 1018–1031. Taylor & Francis.

Lee, S.J., Steiner, R.J., Luo, S., Neale, M.C., Styner, M., Zhu, H., Gilmore, J.H., 2015. Quantitative tract-based white matter heritability in twin neonates. Neuroimage 111, 123–135.

Lin, A., Ching, C.R.K., Vajdi, A., Sun, D., Jonas, R.K., Jalbrzikowski, M., Kushan-Wells, L., et al., 2017. Mapping 22q11.2 gene dosage effects on brain morphometry. J. Neurosci. 37 (26), 6183–6199. Soc Neuroscience.

Lipsky, R.H., Sparling, M.B., Ryan, L.M., Ke, X., Salazar, A.M., Goldman, D., Warden, D.L., 2005. Association of COMT Val158Met genotype with executive functioning following traumatic brain injury. J. Neuropsychiatry Clin. Neurosci. 17 (4), 465–471.

Lipsky, R.H., Lin, M., 2015. Genetic predictors of outcome following traumatic brain injury. Handb. Clin. Neurol. 127, 23–41.

Liu, B., Song, M., Li, J., Liu, Y., Li, K., Yu, C., Jiang, T., 2010. Prefrontal-related functional connectivities within the default network are modulated by COMT val158met in healthy young adults. J. Neurosci. Off. J. Soc. Neurosci. 30 (1), 64–69.

Lukoshe, A., van den Bosch, G.E., van der Lugt, A., Kushner, S.A., Hokken-Koelega, A.C., White, T., 2017. Aberrant white matter microstructure in children and adolescents with the subtype of Prader–Willi syndrome at high risk for psychosis. Schizophr. Bull. 43 (5), 1090–1099. Oxford University Press.

Maas, A.I.R., Menon, D.K., Steyerberg, E.W., Citerio, G., Lecky, F., Manley, G.T., Hill, S., Legrand, V., Sorgner, A., CENTER-TBI Participants and Investigators, 2015. Collaborative European NeuroTrauma Effectiveness Research in Traumatic Brain Injury (CENTER-TBI): a prospective longitudinal observational study. Neurosurgery 76 (1), 67–80. academic.oup.com.

Mac Donald, C.L., Johnson, A.M., Cooper, D., Nelson, E.C., Werner, N.J., Shimony, J.S., Snyder, A.Z., et al., 2011. Detection of blast-related traumatic brain injury in U.S. military personnel. N. Engl. J. Med. 364 (22), 2091–2100.

MacDonald, M.E., Ambrose, C.M., Duyao, M.P., Myers, R.H., Lin, C., Srinidhi, L., Barnes, G., et al., 1993. A novel gene containing a trinucleotide repeat that is expanded and unstable on Huntington's disease chromosomes. Cell 72 (6), 971–983. Elsevier.

Magnotta, V.A., Kim, J., Koscik, T., Beglinger, L.J., Espinso, D., Langbehn, D., Nopoulos, P., Paulsen, J.S., 2009. Diffusion tensor imaging in preclinical Huntington's disease. Brain Imaging Behav. 3 (1), 77–84.

Matsui, J.T., Vaidya, J.G., Johnson, H.J., Magnotta, V.A., Long, J.D., Mills, J.A., Lowe, M.J., et al., 2014. Diffusion weighted imaging of prefrontal cortex in prodromal Huntington's disease. Hum. Brain Mapp. 35 (4), 1562–1573.

Matsui, J.T., Vaidya, J.G., Wassermann, D., Kim, R.E., Magnotta, V.A., Johnson, H.J., Paulsen, J.S., PREDICT-HD Investigators and Coordinators of the Huntington Study Group, 2015. Prefrontal cortex white matter tracts in prodromal Huntington disease. Hum. Brain Mapp. 36 (10), 3717–3732.

Mattiaccio, L.M., Coman, I.L., Schreiner, M.J., Antshel, K.M., Fremont, W.P., Bearden, C.E., Kates, W.R., 2016. Atypical functional connectivity in resting-state networks of individuals with 22q11.2 deletion syndrome: associations with neurocognitive and psychiatric functioning. J. Neurodev. Disord. 8 (1), 2. jneurodevdisorders.biomedcentral.

Mattiaccio, L.M., Coman, I.L., Thompson, C.A., Fremont, W.P., Antshel, K.M., Kates, W.R., 2018. Frontal dysconnectivity in 22q11.2 deletion syndrome: an atlas-based functional connectivity analysis. Behav. Brain Funct. 14 (1), 2. behavioralandbrainfunctions.

McAllister, T.W., Harker Rhodes, C., Flashman, L.A., McDonald, B.C., Belloni, D., Saykin, A.J., 2005. Effect of the dopamine D2 receptor T allele on response latency after mild traumatic brain injury. Am. J. Psychiatry 162 (9), 1749–1751.

McAllister, T.W., Tyler, A.L., Flashman, L.A., Harker Rhodes, C., McDonald, B.C., Saykin, A.J., Tosteson, T.D., Tsongalis, G.J., Moore, J.H., 2012. Polymorphisms in the brain-derived neurotrophic factor gene influence memory and processing speed one month after brain injury. J. Neurotrauma 29 (6), 1111–1118.

McColgan, P., Gregory, S., Razi, A., Seunarine, K.K., Gargouri, F., Durr, A., Roos, R.A.C., et al., 2017. White matter predicts functional connectivity in premanifest Huntington's disease. Ann. Clin. Transl. Neurol. 4 (2), 106–118. Wiley Online Library.

Medland, S.E., Jahanshad, N., Neale, B.M., Thompson, P.M., 2014. Whole-genome analyses of whole-brain data: working within an expanded search space. Nat. Neurosci. 17 (6), 791–800.

Molko, N., Molko, N., Cachia, A., Riviere, D., Mangin, J.F., Bruandet, M., LeBihan, D., Cohen, L., Dehaene, S., 2004. Brain anatomy in turner syndrome: evidence for impaired social and spatial-numerical networks. Cereb. Cortex 14, 840–850.

Moran, L.M., Gerry Taylor, H., Ganesalingam, K., Gastier-Foster, J.M., Frick, J., Bangert, B., Dietrich, A., et al., 2009. Apolipoprotein E4 as a predictor of outcomes in pediatric mild traumatic brain injury. J. Neurotrauma 26 (9), 1489–1495.

Moretti, L., Cristofori, I., Weaver, S.M., Chau, A., Portelli, J.N., Grafman, J., 2012. Cognitive decline in older adults with a history of traumatic brain injury. Lancet Neurol. 11 (12), 1103–1112.

Nadel, L., 1999. Down syndrome in cognitive neuroscience perspective. In: Neurodevelopmental Disorders. MIT Press, Cambridge, MA, pp. 197–222.

Nakamura, T., Hillary, F.G., Biswal, B.B., 2009. Resting network plasticity following brain injury. PLoS One 4 (12), e8220.

Newsome, M.R., Scheibel, R.S., Mayer, A.R., Chu, Z.D., Wilde, E.A., Hanten, G., Steinberg, J.L., et al., 2013. How functional connectivity between emotion regulation structures can be disrupted: preliminary evidence from adolescents with moderate to severe traumatic brain injury. J. Int. Neuropsychol. Soc. 19 (8), 911–924.

Novak, M.J.U., Seunarine, K.K., Gibbard, C.R., Hobbs, N.Z., Scahill, R.I., Clark, C.A., Tabrizi, S.J., 2014. White matter integrity in premanifest and early Huntington's disease is related to caudate loss and disease progression. Cortex 52, 98–112.

Novak, M.J.U., Seunarine, K.K., Gibbard, C.R., McColgan, P., Draganski, B., Friston, K., Clark, C.A., Tabrizi, S.J., 2015. Basal ganglia-cortical structural connectivity in Huntington's disease. Hum. Brain Mapp. 36 (5), 1728–1740. Wiley Online Library.

Nuninga, J.O., Bohlken, M.M., Koops, S., Fiksinski, A.M., Mandl, R.C.W., Breetvelt, E.J., Duijff, S.N., Kahn, R.S., Sommer, I.E.C., Vorstman, J.A.S., 2017. White matter abnormalities in 22q11.2 deletion syndrome patients showing cognitive decline. Psychol. Med. 1–9.

Olszewski, A.K., Kikinis, Z., Gonzalez, C.S., Coman, I.L., Makris, N., Gong, X., Rathi, Y., et al., 2017. The social brain network in 22q11.2 deletion syndrome: a diffusion tensor imaging study. Behav. Brain Funct. 13 (1), 4. behavioralandbrainfunctions.

Padula, M.C., Schaer, M., Scariati, E., Schneider, M., Van De Ville, D., Debbané, M., Eliez, S., 2015. Structural and functional connectivity in the default mode network in 22q11.2 deletion syndrome. J. Neurodev. Disord. 7 (1), 23. jneurodevdisorders.biomedcentral.

Padula, M.C., Schaer, M., Scariati, E., Maeder, J., Schneider, M., Eliez, S., 2017. Multimodal investigation of triple network connectivity in patients with 22q11DS and association with executive functions. Hum. Brain Mapp. 38 (4), 2177–2189. Wiley Online Library.

Pandit, A.S., Expert, P., Lambiotte, R., Bonnelle, V., Leech, R., Turkheimer, F.E., Sharp, D.J., 2013. Traumatic brain injury impairs small-world topology. Neurol. Neurochir. Pol. 80 (20), 1826–1833.

Phillips, O., Sanchez-Castaneda, C., Elifani, F., Maglione, V., Di Pardo, A., Caltagirone, C., Squitieri, F., Sabatini, U., Di Paola, M., 2013. Tractography of the Corpus callosum in Huntington's disease. PLoS One 8 (9), e73280.

Phillips, O.R., Joshi, S.H., Squitieri, F., Sanchez-Castaneda, C., Narr, K., Shattuck, D.W., Caltagirone, C., Sabatini, U., Di Paola, M., 2016. Major superficial white matter abnormalities in Huntington's disease. Front. Neurosci. 10, 197.

Poudel, G.R., Egan, G.F., Churchyard, A., Chua, P., Stout, J.C., Georgiou-Karistianis, N., 2014a. Abnormal synchrony of resting state networks in Premanifest and symptomatic Huntington disease: the IMAGE-HD study. J. Psychiatry Neurosci. 39 (2), 87–96. ncbi.nlm.nih.gov.

Poudel, G.R., Stout, J.C., Domínguez, J.F., Salmon, L., Churchyard, A., Chua, P., Georgiou-Karistianis, N., Egan, G.F., 2014b. White matter connectivity reflects clinical and cognitive status in Huntington's disease. Neurobiol. Dis. 65, 180–187.

Poudel, G.R., Stout, J.C., Domínguez, J.F., Churchyard, A., Chua, P., Egan, G.F., Georgiou-Karistianis, N., 2015. Longitudinal change in white matter microstructure in Huntington's disease: the IMAGE-HD study. Neurobiol. Dis. 74, 406–412.

Powell, D., Caban-Holt, A., Jicha, G., Robertson, W., Davis, R., Gold, B.T., Schmitt, F.A., Head, E., 2014. Frontal white matter integrity in adults with down syndrome with and without dementia. Neurobiol. Aging 35 (7), 1562–1569.

Pujol, J., del Hoyo, L., Blanco-Hinojo, L., de Sola, S., Macià, D., Martínez-Vilavella, G., Amor, M., et al., 2015. Anomalous brain functional connectivity contributing to poor adaptive behavior in down syndrome. Cortex 64, 148–156. Elsevier.

Pujol, J., Blanco-Hinojo, L., Esteba-Castillo, S., Caixàs, A., Harrison, B.J., Bueno, M., Deus, J., et al., 2016. Anomalous basal ganglia connectivity and obsessive-compulsive behaviour in patients with Prader Willi syndrome. J. Psychiatry Neurosci. 41 (4), 261–271.

Quarantelli, M., Salvatore, E., Delle Acque Giorgio, S.M., Filla, A., Cervo, A., Valeria Russo, C., Cocozza, S., Massarelli, M., Brunetti, A., De Michele, G., 2013. Default-mode network changes in Huntington's disease: an integrated MRI study of functional connectivity and morphometry. PLoS One 8 (8), e72159. journals.plos.org.

Rice, L.J., Lagopoulos, J., Brammer, M., Einfeld, S.L., 2017. Microstructural white matter tract alteration in Prader-Willi syndrome: a diffusion tensor imaging study. Am. J. Med. Genet. C Semin. Med. Genet. 175 (3), 362–367.

Risen, S.R., Barber, A.D., Mostofsky, S.H., Suskauer, S.J., 2015. Altered functional connectivity in children with mild to moderate TBI relates to motor control. J. Pediatr. Rehabil. Med. 8 (4), 309–319.

Robinson, M.E., Lindemer, E.R., Fonda, J.R., Milberg, W.P., McGlinchey, R.E., Salat, D.H., 2015. Close-range blast exposure is associated with altered functional connectivity in veterans independent of concussion symptoms at time of exposure. Hum. Brain Mapp. 36 (3), 911–922.

Rosas, H.D., Tuch, D.S., Hevelone, N.D., Zaleta, A.K., Vangel, M., Hersch, S.M., Salat, D.H., 2006. Diffusion tensor imaging in Presymptomatic and early Huntington's disease: selective white matter pathology and its relationship to clinical measures. Movem. Disord. 21 (9), 1317–1325.

Rosas, H.D., Lee, S.Y., Bender, A.C., Zaleta, A.K., Vangel, M., Peng, Y., Fischl, B., et al., 2010. Altered white matter microstructure in the Corpus callosum in Huntington's disease: implications for cortical 'disconnection. Neuroimage 49 (4), 2995–3004.

Rudie, J.D., Hernandez, L.M., Brown, J.A., Beck-Pancer, D., Colich, N.L., Gorrindo, P., Thompson, P.M., et al., 2012. Autism-associated promoter variant in MET impacts functional and structural brain networks. Neuron 75 (5), 904–915.

Sampaio, A., Moreira, P.S., Osório, A., Magalhães, R., Vasconcelos, C., Férnandez, M., Carracedo, A., Alegria, J., Gonçalves, Ó.F., Soares, J.M., 2016. Altered functional connectivity of the default mode network in williams syndrome: a multimodal approach. Dev. Sci. 19 (4), 686–695. Wiley Online Library.

Satizabal, C.L., Adams, H.H.H., Hibar, D.P., White, C.C., 2017. Genetic Architecture of Subcortical Brain Structures in over 40,000 Individuals Worldwide. bioRxiv. biorxiv.org. https://www.biorxiv.org/content/early/2017/08/28/173831.abstract.

Schreiner, M.J., Karlsgodt, K.H., Uddin, L.Q., Chow, C., Congdon, E., Jalbrzikowski, M., Bearden, C.E., 2014. Default mode network connectivity and reciprocal social behavior in 22q11.2 deletion syndrome. Soc. Cogn. Affect. Neurosci. 9 (9), 1261–1267. academic.oup.com.

Schutte, N.M., Hansell, N.K., de Geus, E.J.C., Martin, N.G., Wright, M.J., Smit, D.J.A., 2013. Heritability of resting state EEG functional connectivity patterns. Twin Res. Hum. Genet. 16 (5), 962–969.

Scott-Van Zeeland, A.A., Abrahams, B.S., Alvarez-Retuerto, A.I., Sonnenblick, L.I., Rudie, J.D., Ghahremani, D., Mumford, J.A., et al., 2010. Altered functional connectivity in frontal lobe circuits is associated with variation in the autism risk gene CNTNAP2. Sci. Transl. Med. 2 (56), 56ra80.

Shaffer, J.J., Ghayoor, A., Long, J.D., Kim, R.E.-Y., Lourens, S., O'Donnell, L.J., Westin, C.-F., et al., 2017. Longitudinal diffusion changes in prodromal and early HD: evidence of white-matter tract deterioration. Hum. Brain Mapp. 38 (3), 1460–1477.

Sharp, D.J., Scott, G., Leech, R., 2014. Network dysfunction after traumatic brain injury. Nat. Rev. Neurol. 10 (3), 156–166.

Shen, K.-K., Rose, S., Fripp, J., McMahon, K.L., de Zubicaray, G.I., Martin, N.G., Thompson, P.M., Wright, M.J., Salvado, O., 2014. Investigating brain connectivity heritability in a twin study using diffusion imaging data. Neuroimage 100, 628–641.

Shenton, M.E., Hamoda, H.M., Schneiderman, J.S., Bouix, S., Pasternak, O., Rathi, Y., Vu, M.A., et al., 2012. A review of magnetic resonance imaging and diffusion tensor imaging findings in mild traumatic brain injury. Brain Imaging Behav. 6 (2), 137–192.

Shprintzen, R.J., 2000. Velo-cardio-facial syndrome: A distinctive behavioral phenotype - Shprintzen - 2000 - mental retardation and developmental disabilities research reviews—Wiley online library. Ment. Retard. Dev. Disabil. Res. Rev. 6, 142–147.

Simon, T.J., Ding, L., Bish, J.P., McDonald-McGinn, D.M., Zackai, E.H., Gee, J., 2005. Volumetric, connective, and morphologic changes in the brains of children with chromosome 22q11.2 deletion syndrome: an integrative study. Neuroimage 25, 169–180.

Simon, T.J., Zhongle, W., Avants, B., Zhang, H., Gee, J.C., Stebbins, G.T., 2008. Atypical cortical connectivity and visuospatial cognitive impairments are related in children with chromosome 22q11.2 deletion syndrome. Behav. Brain Funct. 4, 25. behavioralandbrainfunctions.

Sinclair, B., Hansell, N.K., Blokland, G.A.M., Martin, N.G., Thompson, P.M., Breakspear, M., de Zubicaray, G.I., Wright, M.J., McMahon, K.L., 2015. Heritability of the network architecture of intrinsic brain functional connectivity. Neuroimage 121, 243–252.

Smit, D.J.A., Stam, C.J., Posthuma, D., Boomsma, D.I., De Geus, E.J.C., 2008. Heritability of 'small-World' networks in the brain: a graph theoretical analysis of resting-state EEG functional connectivity. Hum. Brain Mapp. 29 (12), 1368–1378. Wiley Online Library.

Smith, S.M., Nichols, T.E., 2018. Statistical challenges in 'Big Data' human neuroimaging. Neuron 97 (2), 263–268. Elsevier.

Sonderby, I., Doan, N.T., Gustafsson, O., Hibar, D., Djurovic, S., Westlye, L.T., Thompson, P., Andreassen, O., ENIGMA-CNV working group, 2017. Association of subcortical brain volumes with CNVS: a mega-analysis from the Enigma-CNV Working Group. Eur. Neuropsychopharmacol. 27, S422–S423. Elsevier.

Spitz, G., Maller, J.J., O'Sullivan, R., Ponsford, J.L., 2013. White matter integrity following traumatic brain injury: the association with severity of injury and cognitive functioning. Brain Topogr. 26 (4), 648–660. Springer.

Stephens, J.A., Salorio, C.E., Gomes, J.P., Nebel, M.B., Mostofsky, S.H., Suskauer, S.J., 2017. Response inhibition deficits and altered motor network connectivity in the chronic phase of pediatric traumatic brain injury. J. Neurotrauma 34 (22), 3117–3123.

Thiruvady, D.R., Georgiou-Karistianis, N., Egan, G.F., Ray, S., Sritharan, A., Farrow, M., Churchyard, A., et al., 2007. Functional connectivity of the prefrontal cortex in Huntington's disease. J. Neurol. Neurosurg. Psychiatry 78 (2), 127–133. jnnp.bmj.com.

Thompson, P.M., Ge, T., Glahn, D.C., Jahanshad, N., Nichols, T.E., 2013. Genetics of the connectome. Neuroimage 80, 475–488.

Trachtenberg, A.J., Filippini, N., Ebmeier, K.P., Smith, S.M., Karpe, F., Mackay, C.E., 2012. The effects of APOE on the functional architecture of the resting brain. Neuroimage 59 (1), 565–572.

Tunbridge, E.M., Farrell, S.M., Harrison, P.J., Mackay, C.E., 2013. Catechol-O-methyltransferase (COMT) influences the connectivity of the prefrontal cortex at rest. Neuroimage 68, 49–54.

Unschuld, P.G., Joel, S.E., Liu, X., Shanahan, M., Margolis, R.L., Biglan, K.M., Bassett, S.S., et al., 2012. Impaired Cortico-Striatal functional connectivity in prodromal Huntington's disease. Neurosci. Lett. 514 (2), 204–209. Elsevier.

van den Heuvel, M.P., van Soelen, I.L.C., Stam, C.J., Kahn, R.S., Boomsma, D.I., Hulshoff Pol, H.E., 2013. Genetic control of functional brain network efficiency in children. Eur. Neuropsychopharmacol. 23 (1), 19–23.

Vega, J.N., Hohman, T.J., Pryweller, J.R., Dykens, E.M., Thornton-Wells, T.A., 2015. Resting-state functional connectivity in individuals with down syndrome and Williams syndrome compared with typically developing controls. Brain Connect. 5 (8), 461–475. online.liebertpub.com.

Verkerk, A.J.M.H., Pieretti, M., Sutcliffe, J.S., Ying-Hui, F., Kuhl, D.P.A., Pizzuti, A., Reiner, O., et al., 1991. Identification of a gene (FMR-1) containing a CGG repeat coincident with a breakpoint cluster region exhibiting length variation in fragile X syndrome. Cell 65, 905–914.

Villalon-Reina, J., Jahanshad, N., Beaton, E., Toga, A.W., Thompson, P.M., Simon, T.J., 2013. White matter microstructural abnormalities in girls with chromosome 22q11.2 deletion syndrome, fragile X or turner syndrome as evidenced by diffusion tensor imaging. Neuroimage 81, 441–454. Elsevier.

Vuoksimaa, E., Panizzon, M.S., Jr, D.J.H., Hatton, S.N., Fennema-Notestine, C., Rinker, D., Eyler, L.T., et al., 2017. Heritability of white matter microstructure in late middle age: a twin study of tract-based fractional anisotropy and absolute diffusivity indices. Hum. Brain Mapp. 38 (4), 2026–2036.

Weaver, K.E., Richards, T.L., Liang, O., Laurino, M.Y., Samii, A., Aylward, E.H., 2009. Longitudinal diffusion tensor imaging in Huntington's disease. Exp. Neurol. 216 (2), 525–529.

Weiner, M.W., Veitch, D.P., Hayes, J., Neylan, T., Grafman, J., Aisen, P.S., Petersen, R.C., et al., 2014. Effects of traumatic brain injury and posttraumatic stress disorder on Alzheimer's disease in veterans, using the Alzheimer's disease neuroimaging initiative. Alzheimers Dement. 10 (3 Suppl), S226–S235.

Wilde, E.A., Ayoub, K.W., Bigler, E.D., Chu, Z.D., Hunter, J.V., Trevor, C.W., McCauley, S.R., Levin, H.S., 2012. Diffusion tensor imaging in moderate-to-severe pediatric traumatic brain injury: Changes within an 18 month post-injury interval. Brain Imaging Behav. 6 (3), 404–416.

Wilde, E.A., Bouix, S., Tate, D.F., Lin, A.P., Newsome, M.R., Taylor, B.A., Stone, J.R., et al., 2015. Advanced neuroimaging applied to veterans and service personnel with traumatic brain injury: state of the art and potential benefits. Brain Imaging Behav. 9 (3), 367–402.

Wolf, R.C., Sambataro, F., Vasic, N., Schönfeldt-Lecuona, C., Ecker, D., Landwehrmeyer, B., 2008. Aberrant connectivity of lateral prefrontal networks in presymptomatic Huntington's disease. Exp. Neurol. 213 (1), 137–144. Elsevier.

Wolf, R.C., Sambataro, F., Vasic, N., Depping, M.S., Thomann, P.A., Landwehrmeyer, G.B., Süssmuth, S.D., Orth, M., 2014. Abnormal resting-state connectivity of motor and cognitive networks in early manifest Huntington's disease. Psychol. Med. 44 (15), 3341–3356. cambridge.org.

Wood, R.L., 2017. Accelerated cognitive aging following severe traumatic brain injury: a review. Brain Inj. 31 (10), 1270–1278.

Xie, S., Yang, J., Zhang, Z., Zhao, C., Bi, Y., Zhao, Q., Pan, H., Gong, G., 2017. The effects of the X Chromosome on intrinsic functional connectivity in the human brain: evidence from turner syndrome patients. Cereb. Cortex 27 (1), 474–484. Oxford University Press.

Xu, M., Yi, Z., von Deneen, K.M., Zhu, H., Gao, J.-H., 2017. Brain structural alterations in obese children with and without Prader-Willi syndrome. Hum. Brain Mapp. 38 (8), 4228–4238.

Yamada, K., Matsuzawa, H., Uchiyama, M., Kwee, I.L., Nakada, T., 2006. Brain developmental abnormalities in Prader-Willi syndrome detected by diffusion tensor imaging. Am Acad Pediatrics 118 (2). e442–48.

Yamagata, B., Barnea-Goraly, N., Marzelli, M.J., Park, Y., Hong, D.S., Mimura, M., Reiss, A.L., 2012. White matter aberrations in Prepubertal estrogen-naive girls with Monosomic turner syndrome. Cereb. Cortex 22 (12), 2761–2768.

Yang, Z., Zuo, X.-N., Katie, L., McMahon, R.C.C., Kelly, C., de Zubicaray, G.I., Hickie, I., et al., 2016. Genetic and environmental contributions to functional connectivity architecture of the human brain. Cereb. Cortex 26 (5), 2341–2352.

Yue, J.K., Vassar, M.J., Lingsma, H.F., Cooper, S.R., Okonkwo, D.O., Valadka, A.B., Gordon, W.A., Maas, A.I.R., Mukherjee, P., Yuh, E.L., 2013. Transforming research and clinical knowledge in traumatic brain injury pilot: Multicenter implementation of the common data elements for traumatic brain injury. J. Neurotrauma 30 (22), 1831–1844.

Yuh, E.L., Cooper, S.R., Mukherjee, P., Yue, J.K., Lingsma, H.F., Gordon, W.A., Valadka, A.B., et al., 2014. Diffusion tensor imaging for outcome prediction in mild traumatic brain injury: a TRACK-TBI study. J. Neurotrauma 31 (17), 1457–1477.

Zhang, Y., Zhao, H., Qiu, S., Tian, J., Wen, X., Miller, J.L., von Deneen, K.M., Zhou, Z., Gold, M.S., Liu, Y., 2013. Altered functional brain networks in Prader-Willi syndrome. NMR Biomed. 26 (6), 622–629. Wiley Online Library.

Zhang, Y., Wang, J., Zhang, G., Zhu, Q., Cai, W., Tian, J., Yi, E.Z., et al., 2015. The neurobiological drive for overeating implicated in Prader–Willi syndrome. Brain Res. 1620, 72–80.

Zhou, W., Di, X., Peng, X., Zhang, Q., Jia, J., Crutcher, K.A., 2008. Meta-analysis of APOE 4 allele and outcome after traumatic brain injury. J. Neurotrauma 25 (4), 279–290.

Characterizing dynamic functional connectivity using data-driven approaches and its application in the diagnosis of alzheimer's disease

Yingying Zhu*, **Xiaofeng Zhu**[†], **Minjeong Kim**[‡], **Daniel Kaufer**[§],
Paul J. Laurienti[¶], **Guorong Wu**[‡]

School of Electrical and Computer Engineer, Cornell University, Ithaca, NY, United States*
*Department of Radiology, Perelman School of Medicine, University of Pennsylvania, Philadelphia,
PA, United States*[†] *Department of Radiology and BRIC, University of North Carolina at Chapel
Hill, Chapel Hill, NC, United States*[‡] *Department of Neurology, University of North Carolina at
Chapel Hill, Chapel Hill, NC, United States*[§] *Department of Radiology, Wake Forest School of
Medicine, Winston-Salem, NC, United States*[¶]

CHAPTER OUTLINE

Functional connectivity (FC) has been widely investigated in many imaging-based neuroscience and clinical studies. Since functional Magnetic Resonance Image (MRI) signal is just an indirect reflection of brain activity, it is difficult to accurately quantify the FC strength only based on signal correlation. To address this limitation,

Connectomics. https://doi.org/10.1016/B978-0-12-813838-0.00010-8

we propose a learning-based tensor model to derive high sensitivity and specificity connectome biomarkers at the individual level from resting-state fMRI images. To model dFC, existing methods for modeling static FC are inefficient for reducing dimensionality and measuring temporal dynamics. For instance, several investigators have tried linear principal component analysis (PCA) to project the high-dimensional dFC onto a low-dimensional spatial connection basis that models the variable connectivity patterns between regions of interest (ROIs). However, such spatial connectivity approaches cannot accurately model the system's temporal dynamics. Most recent work has proposed a robust tensor analysis-based dFC model that reduces the highly dimensional dFC using a spatial connectome scheme in conjunction with temporal dynamics. This model is able to capture the spatial connectivity patterns between ROIs and integrate them with temporal dynamics. Thus, it has the potential to significantly improve brain disorder identification relative to conventional methods.

In this chapter, we will first briefly introduce background of dFC analysis and its application to Alzheimer's disease. Then we will describe the tensor-based statistic model for dFC analysis, which includes three parts: (1) *A data-driven approach to measure FC across brain regions.* Due to low signal-to-noise ratio induced by possible nonneural noise, it is difficult to define an optimal threshold to remove spurious connections. A robust learning-based framework is developed by introducing two regularization terms: sparse constraints on low-level region-to-region signal correlations, as the brain network is intrinsically sparse and low-rank constraints to find the group/modular structure across ROIs. (2) *A data-driven approach to characterize dFC across time.* This is essential to synchronizing the estimated dFC patterns and real-time cognitive state changes. To address these limitations, the previously mentioned FC optimization method is extended to the spatial-temporal domain by arranging the FC estimations using sliding windows into a 3D tensor (third dimension for time). We apply the low-rank constraint on the temporal domain to constrain the dynamic brain network changes smoothly along time. (3) *A statistic model to represent the intrinsic intersubject variation patterns of dynamic functional connectivity.* To characterize the spatial and temporal patterns in high dimensional dFC, two subspaces are created: a spatial FC connection space encoding the variations of region-to-region connections and a temporal FC dynamic space encoding the variations of temporal FC changes across individual subjects. The learned tensor connectome representation is more powerful than the conventional PCA model because the tensor model is able to adaptively capture the spatial FC connection patterns and temporal FC dynamics. The power of the tensor statistical model of dFC is shown in the application to identifying Alzheimer's disease.

INTRODUCTION
DYNAMIC FUNCTIONAL CONNECTIVITY

In general, resting-state FC analyses evaluate the pair-wise associations between the time-series from specific brain areas (Biswal, 2012; Greicius, 2008; Biswal et al., 1995). Each association measure describes the strength of coactivity between the two brain regions. Resting-state FC has been used to identify abnormal brain connectivity patterns in patient cohorts and to understand mechanisms of normal and abnormal brain function. Although FC has not yet directly impacted clinical practice, there is great promise that this technique will yield important biomarkers for disease diagnosis and monitoring in various clinical applications such as Alzheimer's disease (AD) (Aisen et al., 2010; Braak and Braak, 1991; Daianu et al., 2015; Davatzikos et al., 2008; Filley, 1995; Gauthier, 2005; Jagust et al., 2010; Petrie et al., 2009; Stern, 2006; Thompson et al., 2007; Westman et al., 2011; Zhang et al., 2011).

In most functional brain network studies, Pearson's correlation of Blood Oxygen Level Dependent (BOLD) signals is widely used to measure the strength of FC between two brain regions (Amaral et al., 2008; Heuvel and Pol, 2010). It is worth noting that such correlation-based connectivity measures are exclusively calculated based on the observed BOLD signals and fixed for the subsequent data analysis. For simplicity, many FC characterization methods assume that connectivity patterns in the brain are static over the course of a resting-state fMRI scan. There is a growing consensus (Hutchison et al., 2013a; Zhu et al., 2016; Zhu et al., 2017) in the neuroimaging field, however, that the spontaneous fluctuations and correlations of signals between two distinct brain regions change across time, even in a resting state. Conventional analyses that perform correlations using the entire time series do not account for temporal variability and are not sensitive to associated connectivity changes. Thus, dynamic FC patterns have been primarily investigated recently by using a sliding-window technique (Hutchison et al., 2013a). However, it is very difficult to synchronize the estimated dynamic patterns with the real fluctuations of cognitive state, even using the advanced machine learning techniques such as unsupervised clustering (Wee et al., 2016) and hidden Markov model (Eavani et al., 2013). For example, both methods have to determine the number of states (clusters) that might work well on the training data, but may not generalize well to the testing subjects.

In this chapter, we will discuss a recently published (Zhu et al., 2017; Wee et al., 2016) robust data-driven solution to reveal the consistent spatial-temporal FC patterns from resting-state fMRI image using a fixed sliding window, as shown in Fig. 1. To find the sparse connectivity between ROIs and the grouped ROIs relationship based on signal correlations, both local sparsity constraints and global grouped ROIs constraints are used to guide the learning process of FC. Specifically, the high-level modular constraints are achieved by applying low-rank constraints on the FC matrix. The low-rank constraints enable groupwise connectivity among ROIs that comprise modular structure in the brain. Because brain networks are intrinsically sparse, this sparsity constraint is used to control the number of connections between all ROIs. Based on this robust data-driven approach, we further introduce a dynamic brain connectivity method, which extend the previous FC optimization framework from one sliding window to a set of overlapped sliding windows. The FCs along time are arranged into a tensor structure (pink cubic in Fig. 1), where the third dimension represents the scanning time. A low-rank constraint is applied to the time domain to limit high-frequency noise from contaminating the FC temporal state changes.

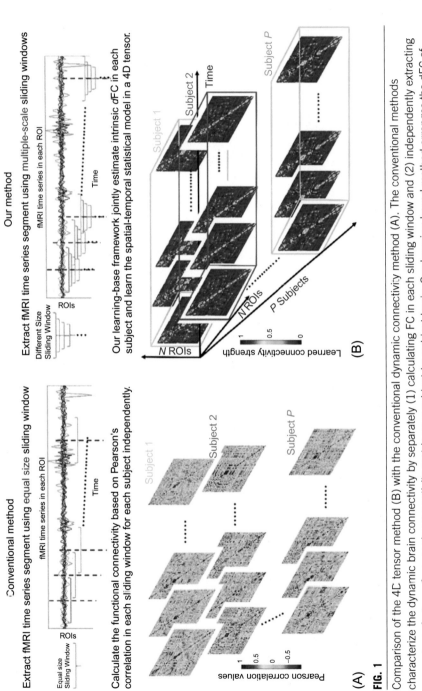

FIG. 1

Comparison of the 4C tensor method (B) with the conventional dynamic connectivity method (A). The conventional methods characterize the dynamic brain connectivity by separately (1) calculating FC in each sliding window and (2) independently extracting connectome features from each connectivity matrix on a subject-by-subject basis. Our learning-based method arranges the dFC of each subject into a 3D tensor and further stacks the subject-specific 3D tensors into the population-wise 4D tensor. Contrary to the conventional methods, our data-driven method jointly optimizes dFC at the individual level and learns the spatial-temporal statistical model at the population level using the 4D tensor analysis.

Furthermore, considering the relationship among subjects, we stack the dFC (in the 3D tensor) of all subjects into a 4D tensor, where the fourth dimension represents the individual subjects in the population. Contrary to the conventional method, this data-driven solution jointly optimizes dFC at the individual level and learns the spatial-temporal statistical model at the population level using the 4D tensor analysis.

ALZHEIMER'S DISEASE

AD, the most common neurodegenerative disorder, leads to gradually progressive memory loss, decline in other cognitive domains, altered behavior, loss of functional abilities, and ultimately death (Filley, 1995; Gauthier, 2005; Jagust et al., 2010; Petrie et al., 2009; Thompson et al., 2007; Rockwood, 2010; Chetelat and Baron, 2003; Schaeffer et al., 2011; Xie and He, 2012). As shown in Fig. 2, AD is a progressive disease such that it starts from the medial temporal lobe and gradually spreads throughout the brain. Convergent evidence shows that (1) deterioration of brain structures commences several years prior to the cognitive decline and (2) such time period has been estimated to be several years or even decades before the onset of AD clinical symptoms (Thompson et al., 2007; Viola et al., 2015). However, current diagnosis of AD is usually made after the initial clinical symptom appears. Because current therapeutic interventions may have the most meaningful clinical effect in the early disease stage, early detection of AD symptoms is very important for optimizing the benefit of AD treatments (Gauthier, 2005; Chetelat and Baron, 2003; Viola et al., 2015; Chincarini et al., 2016; Cummings et al., 2007; Geldmacher et al., 2014; Nestor et al., 2004; Weimer and Sager, 2009).

AD is also a complex neurodegenerative disease that manifests in different symptoms. Although progressive neuron loss is a hallmark of CEAD (Schaeffer et al., 2011; Serran-Pozo et al., 2011; Morrison and Hof, 1997), some neurological impairment may reflect dysfunction rather than loss of neurons (Palop et al., 2006). For example, many AD patients display remarkable fluctuations in neurological function across time. These fluctuations most likely reflect variations in neural network activity potentially caused by abnormal protein aggregations. Such aggregations have shown to trigger cycles of aberrant neuronal activity and compensatory alternations in neurotransmitter receptors. Repeated cycles can lead to synaptic deficits, the

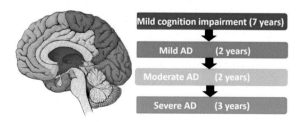

FIG. 2

The progression of Alzheimer's disease.

disintegration of the neural network, and ultimately failure of neurological function (Palop et al., 2006; Arendt, 2003). In this light, AD can be understood as a disconnection syndrome where the structural and functional connectivity of the large-scale network is progressively disrupted by neuropathological processes that are not fully understood (Gomez-Ramirez and Wu, 2014; Seeley et al., 2009; Stam, 2014).

CONSTRUCTING ROBUST STATIC FUNCTIONAL CONNECTIVITY

DATA-DRIVEN APPROACH TO MEASURE FUNCTIONAL CONNECTIVITY

Whole-brain FC can be encoded in a $N \times N$ symmetric matrix S where each element s_{ij} indicates the strength of connection between region O_i and O_j ($i, j = 1, ..., N$). We have developed a data-driven FC analysis constrained by three well-established properties of brain network connectivity: (1) connectivity should be similar with the Pearson's correlation of low-level signals between x_i and x_j; (2) the connectivity matrices should be sparse because the brain network is intrinsically efficient with sparse connections (Bullmore and Sporns, 2009); and (3) the estimated connectivity matrix should exhibit modular organization (Sporns, 2013; Rubinov and Sporns, 2010).

For convenience, we use $s_i \in \Re^{N \times 1}$ to denote i-th column in connectivity matrix S, which characterizes the connections of region O_i with respect to other brain regions. Also, we arrange all Pearson's correlation values into another $N \times N$ matrix $C = \{c_{ij} | j = 1, \cdots, N\}$. Instead of calculating the connectivity s_{ij} just based on Pearson's correlation $c(x_i, x_j)$ between observed BOLD signals x_i and x_j, the connectivity matrix S is learned by integrating the previous two criteria:

$$\text{argmin}_S \| S - C \|_F^2 + \alpha \| S \|_* + \beta \| S \|_1 \tag{1}$$

where α and β are the scalars that balance the strength of the low-rank constraint (Barranca et al., 2015) on S (the second term) and the sparse constraint (Nie et al., 2010) on S (the third term).

The first term in Eq. (1) requires that the optimized FC matrix S should be close to the observed Pearson's correlation matrix C. The intuition of sparsity and low rank constraints are demonstrated in the toy example in Fig. 3. Fig. 3A shows a typical FC matrix calculated based on Pearson's correlation. It is usually difficult to remove the spurious connections based on single threshold, resulting in too many connections at each node. After applying the sparse constraint independently at each brain region (set $\alpha = 0$ in Eq. (1)), only a small number of connections remain such that the number of real versus spurious connections is increased, resulting in more robust estimation of brain network, as shown in Fig. 3B. It is worth noting that sparse representation technique has been used by Lv and Wee (Lv et al., 2015; Wee et al., 2014) to establish the functional connection of one brain region to all other regions by representing the mean time course using the other mean time course signals from other regions. The values of the sparse representation coefficients are proportional to the strength of functional connectivity.

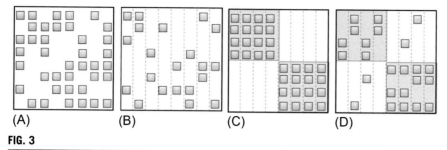

FIG. 3

The intuition behind low-rank and sparsity constraints in optimizing functional connectivity. (A) Typical FC matrix by using Pearson's correlation and thresholding technique that likely contains many spurious connections. (B) Sparse constraints achieve sparse inner-module and cross-module connections equally, which has weak modular organization. (C) Low-rank constraints can produce strong modular organization, however, inner-module regions are overconnected. (D) Balance sparse constraints and low-rank constraints leads to reasonable connectivity patterns, which shows modular organization and sparse inner-module and inter-module connections.

The L_1 sparsity constraint effectively reduces the uncertain and redundant connections in the functional network. However, one limitation of using L_1 sparse constraint is that the modular relationship in S is not jointly considered. There is no guarantee that the resultant sparse network will exhibit modular organization comparable to that known to exist in the human connectome (Heuvel and Sporns, 2011). Because brain networks are known to exhibit modular organization, we further introduce a low-rank constraint $\|S\|_*$ to entire connectivity matrix S, which produces clear modular organization. Fig. 3C shows the optimized connectivity matrix only using low-rank constraint (i.e., $\beta = 0$). It is clear that the strong low-rank constraint can yield connection patterns that are overly organized such that there are only within module connections and not between module connections. Applying a threshold to the modular network at this point would still result in a network with excessive within-module connections and few if any between-module connections. Such a network would be fragmented, and each module would be an isolated, completely connected component. To achieve the desired sparse and modular brain network, we propose to combine low-rank and sparsity constraint on S, with the desired optimization result shown in Fig. 3D.

DATA-DRIVEN APPROACH TO CHARACTERIZE DYNAMIC FUNCTIONAL CONNECTIVITIES

In this section, we extend the learning-based FC optimization method to the temporal domain to model FC dynamics (Zhu et al., 2016). First, we follow the sliding-window technique to obtain T overlapped multiscale sliding windows. Those sliding windows cover the whole time course for each subject, as shown in the top of Fig. 1B. Let S^t denote the FC matrix in sliding window t. Then we stack all S^t along time and form a 3D tensor $\mathcal{S} = \left[\mathbf{S}^t \right]_{t=1,...,T}$ ($\mathcal{S} \in \mathfrak{R}^{N \times N \times T}$), which represents the dynamic connectivity of networks for each subject. Similarly, we can also construct another 3D tensor $\mathcal{C} = \left[\mathbf{C}^t \right]_{t=1,...,T}$ ($\mathcal{C} \in \mathfrak{R}^{N \times N \times T}$), where each $\mathbf{C}^t = [c^t(i,j)]_{i,j=1,...,N}$ is the $N \times N$ matrix in the t-th sliding window. Next, we propose a learning-based optimization method to characterize dFC using tensor analysis by:

$$\operatorname{argmin}_{\mathcal{S}} \| \mathcal{S} - \mathcal{C} \|_{\mathrm{F}}^2 + \alpha \| \mathcal{S} \|_* + \beta \| \mathcal{S} \|_1 \tag{2}$$

Because the brain generally transverses a small number of discrete stages during a short period of time in resting state (Hutchison et al., 2013b), it is reasonable to apply low-rank constraint on the 3D tensor $\mathcal{S}$ (by minimizing nuclear norm $\mathcal{S}_*$). This not only encourages modular organization within each sliding window (in the spatial domain) but also suppresses many rapid connectivity changes across sliding windows (in the temporal domain).

STATISTIC MODEL TO CAPTURE FUNCTIONAL CONNECTIVITY VARIATIONS

The previously descibed work only focused on the optimization of dFC $\mathcal{S}_p$ for each individual subject I_p. Here we use variable p to denote the index of subject. It is apparent that $\mathcal{S}_p$, is encoded in a $N \times N \times T$ 3D tensor, and has far too many data dimensions and excessive redundancy. Hence, it is not common to directly use $\mathcal{S}_p$ to train a classifier for disease diagnosis. Inspired by the statistical models widely used in computer vision that focus on shape and morphological patterns (Cootes et al., 2001; Cootes et al., 1995), the existing dimension reduction approaches employ PCA to capture major variations among either a set of vectorized dFCs or FCs. For the former, the training data consists of P column vectors, each with ($N^2 \cdot T$) long, vectorized from the dFC of P subjects. With this method, the dynamic connectivity information is mixed with the spatial connectivity information within each sliding window. Unfortunately, the relatively large variations of FC across individual subjects tend to dominate the more subtle intrasubject patterns of functional dynamics in the learned statistical model. For the latter, the dFC of each subject is first broken down into T column vectors where each column vector is a N^2 long column vector derived from the $N \times N$ connectivity matrix in a particular sliding window. Then, PCA is applied to the total ($P \cdot T$) column vectors to learn the statistical model, despite that two column vectors are from the same subject but different sliding windows or the same window but different subjects. It is apparent that, with this method, the dynamic information is not well maintained, a critical issue as there is growing evidence that functional dynamics are of great importance in disease diagnosis.

To fully utilize the information of functional dynamics, we extend our previous work from 3D tensor to 4D tensor analysis as follows: Suppose we have P training subjects in total. First, we construct a 4D tensor $\mathbb{C} = \left[\mathcal{C}_p \right]_{p=1,...,P}$ ($\mathbb{C} \in \mathfrak{R}^{N \times N \times T \times P}$), where $\mathcal{C}_p = \left[\mathbf{C}_p^t \right]_{t=1,...,T}$ is the 3D tensor of dFC measures by Pearson's correlation. Similarly, we construct the 4D tensor $\mathbb{S} = \left[\mathcal{S}_p \right]_{p=1,...,P}$ ($\mathbb{S} \in \mathfrak{R}^{N \times N \times T \times P}$) where each $\mathcal{S}_p = \left[\mathbf{S}_p^t \right]_{t=1,...,T}$ is the to-be-optimized dFC formed in the 3D tensor. Next, we propose to iteratively optimize $\mathbb{S}$ for all P training subjects and build the statistical model of dFC.

First, the 4D tensor $\mathbb{S}$ is constrained to: (1) be close to the observed 4D tensor $\mathbb{C}$ measured by Pearson's correlation (i.e., $\|\mathbb{S}-\mathbb{C}\|_2^2 \to 0$); (2) have sparse connectivity among brain regions; and (3) have the property of low rank, which presents modularity within each connectivity matrix S_p^t, consistency along acquisition time within the dFC $\mathcal{S}_p$ of each subject, and less redundancy across all individuals in $\mathbb{S}$. Furthermore, we consider the intersubject variations in $\mathbb{S}$ consisting of two aspects: (1) the spatial patterns of FC across individuals and (2) the dynamic patterns of connectivity changes in the population. We learn the basis of spatial connections and temporal dynamics simultaneously such that the tentatively optimized $\mathbb{S}$ can be reconstructed from the projections in these two bases. Specifically, we reshape the 4D tensor $\mathbb{S}$ into a $N^2 \times T \times P$ 3D tensor, denoted by $reshape_{4D \to 3D}(\mathbb{S})$, because the $N \times N$ connectivity matrix is symmetric. The basis vectors $U_c \in \mathfrak{R}^{N^2 \times R}$ and $U_t \in \mathfrak{R}^{T \times R}$ form the spatial connection basis (regarding the region-to-region functional connectivity) and temporal dynamic basis (regarding the temporal connection changes of each ROI), respectively. Thus, the low-dimension tensor connectome codes $\mathbb{F} = [F_p]_{p=1,...,P}$ ($\mathbb{F} \in \mathfrak{R}^{R \times R \times P}$) are computed from the reshaped dFCs $\mathbb{S}$ by $\mathbb{F} = reshape_{4D \to 3D}(\mathbb{S}) \times_1 U_{(c)} \times_2 U_{(t)}$, where $\times_1$ and $\times_2$ represent the tensor mode multiplication. Similarly, given the low-dimension tensor connectome codes $\mathbb{F}$, we can reconstruct the dFCs by $reshape_{4D \to 3D}(\mathbb{S}) = \mathbb{F} \times_1 U_{(c)} \times_2 U_{(t)}$. Because we reduce the dimension of statistical patterns from N^2 to R ($R \ll N^2$) and reduce the dimension of dynamic patterns from T to R ($R \ll T$), we encourage the reconstruction result $\mathbb{F} \times_1 U_{(c)} \times_2 U_{(t)}$ to approximate the original dFCs $\mathbb{S}$ (i.e., $\|reshape_{4D \to 3D}(\mathbb{S}) - \mathbb{F} \times_1 U_{(c)} \times_2 U_{(t)}\|_2^2 \to 0$). It is apparent that our tensor connectome model, denoted as $\mathcal{M}$, eventually consists of the basis vectors $U_{(c)}$ and $U_{(t)}$, i.e., $\mathcal{M} = \{U_{(c)}, U_{(t)}\}$. The objective function of our joint optimization of dFC $\mathbb{S}$ and learning of statistical model $\mathcal{M}$ is defined as:

$$\arg\min_{\mathbb{F},U_{(c)},U_{(t)},\mathbb{S}} \left\{ \|\mathbb{S}-\mathbb{C}\|_2^2 + \alpha\|\mathbb{S}\|_* + \beta\|\mathbb{S}\|_{2,1} + \gamma\|\mathbb{S} - reshape_{3D \to 4D}(\mathbb{F} \times_1 U_{(c)} \times_2 U_{(t)})\|_2^2 \right\} \quad (3)$$

where α, β, and γ are scalars balancing the effect of each constraint. As we will explain next, the low-dimension tensor connectome codes $\mathbb{F}$ for P training subjects are the key factor for achieving joint optimization of $\mathbb{S}$ and the learning of the statistical model $\mathcal{M}$.

OPTIMIZATION OF TENSOR STATISTIC MODEL

The objective function in Eq. (3) can be optimized by alternatively solving the following two subproblems:

Subproblem 1. Optimize the dFC for all training subjects under the guidance of the tensor connectome model. By fixing the model parameters $\mathcal{M}$ and the low-dimension tensor connectome codes $\mathbb{F}$, we can optimize dFC $\mathbb{S}$ by:

$$\arg\min_{\mathbb{S}} \|\mathbb{S}-\mathbb{C}\|_2^2 + \alpha\|\mathbb{S}\|_* + \beta\|\mathbb{S}\|_{2,1} + \gamma\|\mathbb{S} - reshape_{3D \to 4D}(\mathbb{F} \times_1 U_{(c)} \times_2 U_{(t)})\|_2^2. \quad (4)$$

It is worth noting that the first and last terms in Eq. (4) are the data fidelity terms regarding the observed FC measured by Pearson's correlation and confirmed using the tentatively learned statistical model.

Subproblem 2. Learn the low-dimensional representation of dynamic functional connectivity. Given the tentatively optimized dFCs $\mathbb{S}$, we first reshape $\mathbb{S}$ to the $N^2 \times T \times P$ 3D tensor and then apply a classic tensor analysis method to find the basis vectors $U_{(c)}$ and $U_{(t)}$ and determine the low-dimension tensor connectome features $\mathbb{F}$:

$$\arg\min_{\mathbb{F}, U_{(c)}, U_{(t)}} \| \mathbb{S} - reshape_{3D \to 4D} (\mathbb{F} \times_1 U_{(c)} \times_2 U_{(t)}) \|_2^2 \tag{5}$$

OBTAIN COMPACT REPRESENTATION BY THE LEARNED TENSOR STATISTIC MODEL

The learned $U_{(c)}$, $U_{(t)}$, and the low-dimension tensor connectome features $\mathbb{F}$ are used to guide the optimization of dFCs in Eq. (4). Subproblems 1 and 2 are solved jointly, and we use Alternating Direction Method of Multipliers (ADMM) to solve them iteratively until they converge.

In the application stage, each testing subject is assumed be independent of the other subjects. Thus, the low-dimension tensor connectome features for each testing subject I_q can be calculated separately. Specifically, the 3D tensor of Pearson's correlation $\mathcal{C}_q \in \mathfrak{R}^{N \times N \times T}$ is calculated first, using T sliding windows. Next, the dFC $\mathcal{S}_q$ ($\mathcal{S}_q \in \mathfrak{R}^{N \times N \times T}$) and tensor connectome feature F_q ($F_q \in \mathfrak{R}^{R \times R}$) for the testing subject are obtimized jointly by:

$$\arg\min_{\mathcal{S}_q, F_q} \| \mathcal{S}_q - \mathcal{C}_q \|_2^2 + \alpha \| \mathcal{S}_q \|_* + \beta \| \mathcal{S}_q \|_{2,1} + \gamma \| \mathcal{S}_q - F_q \times_1 U_{(c)} \times_2 U_{(t)} \|_2^2 \tag{6}$$

The similar optimization procedure in "Optimization of Tensor Statistic Model" section. are used here by fixing the basis vectors $U_{(c)}$ and $U_{(t)}$. They are learned from the training dataset. In brief, we solve the optimization in Eq. (6) by alternating the following two steps: (1) estimate the dFC $\mathcal{S}_q$, which should be close to both $\mathcal{C}_q$ and the reconstructed 3D tensor based on the low-dimension ($R \times R$) code F_q; and (2) refine the low-dimension tensor connectome code F_q by projecting to the learned connectome basis $U_{(c)}$ and dynamic basis $U_{(t)}$. In the end, the optimized low-dimension tensor code F_q is regarded as the connectome signature of testing subject I_q. The classification result for testing subject I_q is determined through a trained support vector machine (SVM) by inputting the low-dimension tensor connectome code F_q.

Specifically, a SVM is used to train the classifier based on the learned low-dimension tensor connectome codes $\mathbb{F}$ for the total P training subjects. It is worth noting that any classifier can be applied to our learned tensor connectome feature. Here, to show the power of tensor connectome features derived from our tensor connectome model, we use the standard kernel SVM with L_2 norm penalty to train the classifier. Specifically, libSVM package is used here as it is one of the most popular SVM solvers in the public domain (https://www.csie.ntu.edu.tw/~cjlin/libsvm/).

APPLICATION OF TENSOR STATISTIC MODEL IN AD DIAGNOSIS

In this section, the optimized dFC and spatial-temporal statistic model of dFC is evaluated on the fMRI images collected from the Alzheimer's Disease Neuroimaging Initiative (ADNI) database (http://adni.loni.usc.edu/) to diagnose AD and mild cognitive impairment (MCI). The results from the tensor connectome model are compared with the recent Eigen connectivity feature model (Leonardi et al., 2013) in terms of discrimination power in diagnosis.

Subject information. The following experiments using the resting-state fMRI images from the ADNI database are conducted. Specifically, 98 normal controls (NC) and 118 MCI participants were selected. And there are 55 early MCI (EMCI) subjects and 63 late MCI (LMCI) subjects in the MCI participants.

Data preprocessing. The participants were scanned for 6 and 10 min during resting state. These scan durations at a repetition time (TR) of 2 s resulted in 180 time points and 300 time points, respectively. All the data were processed using Data Processing Assistant for Resting-State fMRI (DPARSF) software. Specifically, the first 20 time points and last 20 time points are removed from each image series. After that, slice timing and motion correction were performed. Then, the Automated Anatomical Labeling (AAL) atlas (Tzourio-Mazoyer et al., 2002) with 90 ROIs was registered to the native brain space of each individual participant using deformable image registration (Wu et al., 2013). Next, the mean BOLD time courses for each ROI were computed. These data were used to calculate the 90×90 Pearson's correlation matrix.

Evaluation of spatial-temporal tensor connectome model. We evaluated the learned spatial-temporal tensor connectome statistical model. Recall that our model consists of spatial connection basis $U_{(c)}$ and the temporal dynamic basis vectors $U_{(t)}$. Specifically $U_{(c)} = [u_{(c)}^r]_{r=1,...R}$ is a $N^2 \times R$ matrix where each column vector $u_{(c)}^r$ can be reshaped back to a $N \times N$ connectivity matrix. Of note, we sort u_r in a decreasing order based on the energy of pattern variation. Fig. 4 shows the top nine principal component connectivity patterns from the population. Of note, these connectivity patterns account for almost 40% energy of the total variations in the population. We partition 90 brain regions into six areas (i.e., frontal lobe, cingulate lobe, temporal lobe, occipital lobe, partial lobe, and subcortical [SBC]). It is apparent that the number of functional connections within the area was much larger than the number of functional connections across areas. For example, the instances of FC in first two largest components are mainly between the regions within subcortical and occipital lobe, respectively. As anticipated, there were a considerable number of connections between hemispheres connecting homologous brain regions.

The temporal dynamic basis vectors $U_{(t)} = [u_{(t)}^r]_{r=1,...,R}$ consists of R column vectors, where each column vector denotes the trajectory of functional connectivity. Similarly, we display the top 11 most frequent dynamic components in Fig. 5, where each dynamic component characterizes the trajectory of FC changes. It is interesting to find that (1) the learned temporal dynamic basis is periodic, as the shape of curves looks like the basis function in Discrete Cosine Transform (DCT); (2) the temporal dynamic basis shows more low-frequency changes than high-frequency changes; and (3) the top temporal dynamic basis is almost stationary, which reflects the nature of resting state.

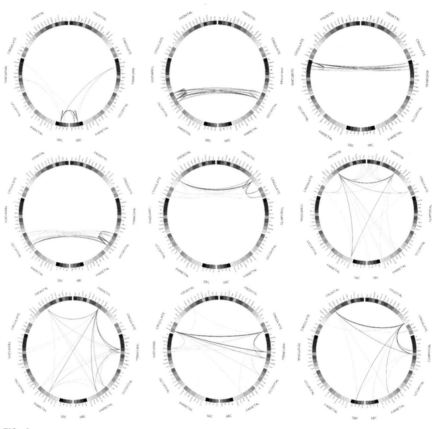

FIG. 4

Visualization of top nine principal components of FC patterns presented learned by our tensor-based statistical model. There are 90 regions on the circular representation graph from the frontal lobe, cingulate lobe, temporal lobe, occipital lobe, partial lobe, and subcortical (SBC). The left hemisphere is on the left side of the circle.

Evaluation of learned tensor connectome features in disease diagnosis. A 10-fold cross validation strategy was used in all the following experiments. Specifically, we randomly partition all subjects into 10 nonoverlapping, approximately equal size sets. Then, we use one fold for testing, and the remaining folds were used for training. The training data were further divided into five subsets for another fivefold inner cross validation, where fourfolds were used as training subsets and the last fold was used as the validation subset. For the competing method (Leonardi et al., 2013), we applied the sliding-window strategy to calculate the dFC feature representation based on Pearson's correlation with optimal threshold determined on a validation dataset. To reduce the feature dimensions, we followed the work in Leonardi et al. (Leonardi et al., 2013) to use a classic PCA model to derive the Eigen spatial connectome basis

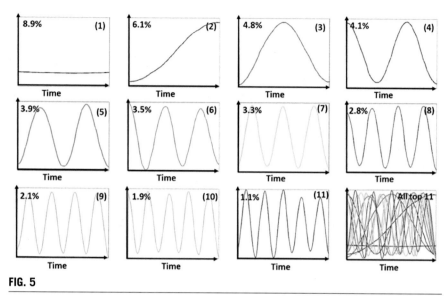

FIG. 5

Visualization of top 11 temporal dynamics basis learned from the tensor connectome model.

and low-dimensional feature representation for each testing subject. We also evaluated the performance of using conventional node degree features for dFC as proposed in Price et al. (Price et al., 2014). To capture both short-term and long-term dynamics, we further integrated the sliding windows with different sizes ranging from 20 to 100 time points. It is worth noting that the integration of multiscale sliding windows is straightforward for our tensor connectome model as the training and testing procedures remain intact but with an increased number of sliding windows T. We used Accuracy (ACC) and Area Under the receiver operating curve (AUC) to evaluate classification performance.

Experiments on ADNI dataset. Recent work has shown high performance when attempting to classify normal control (NC) older adults from those with AD. In this work, we also applied our model to more clinically relevant and challenging classification tasks: AD vs NC, AD vs MCI, MCI vs NC, and early MCI (EMCI) vs late MCI (LMCI) classification.

The AD/NC, AD/MCI, MCI/NC, and EMCI/LMCI classification results using different sliding-window sizes are shown in Figs. 6–9, respectively. It is apparent that the multiscale sliding-window method achieves the best performance for both the PCA model and the proposed tensor connectome methods. The performance improvements were >2% compared with the best performance achieved by fixed sliding-window size. Also, our tensor connectome model exhibited better performance by at least 4.0% compared with the PCA Eigen connectivity model, even using multiscale sliding windows.

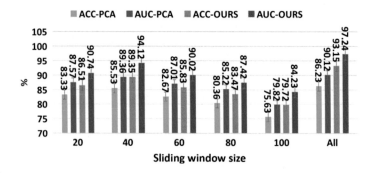

FIG. 6

Accuracy of identifying AD subjects from NC subjects w.r.t. sliding-window size for the current tensor model (OURS) and the classic PCA model (PC).

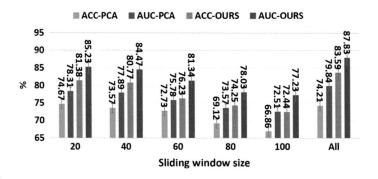

FIG. 7

Accuracy of identifying AD subjects from MCI subjects w.r.t. sliding-window size for the current tensor model (OURS) and the classic PCA model (PC).

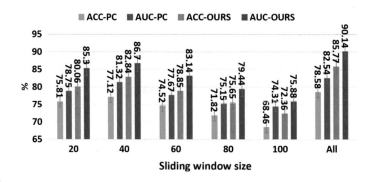

FIG. 8

Accuracy of identifying MCI subjects from NC subjects w.r.t. sliding-window size for the current tensor model (OURS) and the classic PCA model (PC).

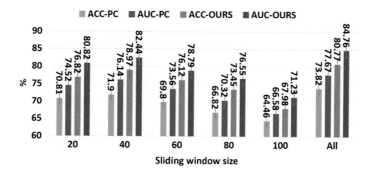

FIG. 9

Accuracy of identifying EMCI subjects from LMCI subjects w.r.t. sliding-window size for the current tensor model (OURS) and the classic PCA model (PC).

CONCLUSION

In this chapter, we introduced a robust statistical model for constructing dynamic functional brain networks. This model uses the tensor analysis approach to jointly optimize the dynamic function connectivity for individuals and also generates a compact tensor code for dFC across a population. The learned compact tensor code can be used as a dFC feature for disease diagnosis. Experiments show that the tensor connectome feature can significantly improve the classification accuracy of MCI and AD compared with current methods that comprise the *status quo*. The promising results in these experiments implicate great potential for our tensor connectome model in clinical applications.

REFERENCES

Aisen, P.S., et al., 2010. Clinical Core of the Alzheimer's disease neuroimaging initiative: progress and plans. Alzheimers Dement. 6, https://doi.org/10.1016/j.jalz.2010.03.006.

Amaral, D.G., Schumann, C.M., Nordahl, C.W., 2008. Neuroanatomy of autism. Trends Neurosci. 31, 137–145.

Arendt, T., 2003. Synaptic plasticity and cell cycle activation in neurons are alternative effector pathways: the 'Dr. Jekyll and Mr. Hyde concept' of Alzheimer's disease or the yin and yang of neuroplasticity. Prog. Neurobiol. 71, 83–248.

Barranca, V.J., Zhou, D., Cai, D., 2015. Low-rank network decomposition reveals structural characteristics of small-world networks. Phys. Rev. E 92 (062822).

Biswal, B.B., 2012. Resting state fMRI: a personal history. Neuroimage 62, 938–944.

Biswal, B., Yetkin, F.Z., Haughton, V.M., Hyde, J.S., 1995. Functional connectivity in the motor cortex of resting human brain using echo-planar MRI. Magn. Reson. Med. 34, 537–541.

Braak, H., Braak, E., 1991. Neuropathological staging of Alzheimer-related changes. Acta Neuropathol. 82, https://doi.org/10.1007/bf00308809.

Bullmore, E., Sporns, O., 2009. *Complex* brain networks: graph theoretical analysis of structural and functional systems. Nature 10, 186–198.

Chetelat, G.a., Baron, J.-C., 2003. Early diagnosis of alzheimer's disease: contribution of structural neuroimaging. Neuroimage 18, 525–541.

Chincarini, A., et al., 2016. Integrating longitudinal information in hippocampal volume measurements for the early detection of Alzheimer's disease. Neuroimage 125, 834–847.

Cootes, T.F., Taylor, C.J., Cooper, D.H., Graham, J., 1995. Active shape models - their training and application. Comput. Vis. Image Underst. 61, 38–59.

Cootes, T.F., Edwards, G.J., Taylor, C.J., 2001. Active appearance models. IEEE Trans. Pattern Anal. Mach. Intell. 23, 681–685.

Cummings, J.L., Doody, R., Clark, C., 2007. Disease-modifying therapies for Alzheimer disease: challenges to early intervention. Neurology 69, https://doi.org/10.1212/01.wnl.0000295996.54210.69.

Daianu, M., et al., 2015. Rich club analysis in the Alzheimer's disease connectome reveals a relatively undisturbed structural core network. Hum. Brain Mapp. 36, 3087–3103.

Davatzikos, C., Fan, Y., Wu, X., Shen, D., Resnick, S.M., 2008. Detection of prodromal Alzheimer's disease via pattern classification of MRI. Neurobiol. Aging 29, 514–523.

Eavani, H., Satterthwaite, T., Gur, R., Gur, R., Davatzikos, C., 2013. Information Processing in Medical Imaging.

Filley, C., 1995. Alzheimer's disease: it's irreversible but not untreatable. Geriatrics 50, 18–23.

Gauthier, S.G., 2005. Alzheimer's disease: the benefits of early treatment. Eur. J. Neurol. 12, 11–16.

Geldmacher, D., et al., 2014. Implications of early treatment among Medicaid patients with Alzheimer's disease. Alzheimers Dement. 10, 214–224.

Gomez-Ramirez, J., Wu, J., 2014. Network-based biomarkers in Alzheimer's disease: review and future directions. Front. Aging Neurosci. 6, .

Greicius, M., 2008. Resting-state functional connectivity in neuropsychiatric disorders. Curr. Opin. Neurol. 21, 430–434.

Heuvel, M.P.v.d., Pol, H.E.H., 2010. Exploring the brain network: a review on resting-state fMRI functional connectivity. Eur. Neuropsychopharmacol. 20, 519–534.

Heuvel, M.P.v.d., Sporns, O., 2011. Rich-club organization of the human connectome. J. Neurosci. 31, 15775–15786.

Hutchison, R.M., et al., 2013a. Dynamic functional connectivity: promise, issues, and interpretations. Neuroimage 80, 360–378.

Hutchison, R.M., et al., 2013b. Dynamic functional connectivity: promise, issues, and interpretations. Neuroimage 80, 360–378.

Jagust, W.J., et al., 2010. The Alzheimer's disease neuroimaging initiative positron emission tomography core. Alzheimers Dement. 6, https://doi.org/10.1016/j.jalz.2010.03.003.

Leonardi, N., et al., 2013. Principal components of functional connectivity: a new approach to study dynamic brain connectivity during rest. Neuroimage 83, 937–950.

Lv, J., et al., 2015. Sparse representation of whole-brain fMRI signals for identification of functional networks. Med. Image Anal. 20, 112–134.

Morrison, J.H., Hof, P.R., 1997. Life and death of neurons in the aging brain. Science 278, 412–419.

Nestor, P.J., Scheltems, P., Hodges, J.R., 2004. Advances in the early detection of Alzheimer's disease. Nat. Rev. Neurosci. 5, S34–S41.

Nie, F., Huang, H., Cai, X., Ding, C., 2010. Advances in Neural Information Processing Systems 23 (NIPS).

Palop, J.J., Chin, J., Mucke, L., 2006. A network dysfunction perspective on neurodegenerative diseases. Nature 443, 768–773.

Petrie, E.C., et al., 2009. Preclinical evidence of Alzheimer changes: convergent cerebrospinal fluid biomarker and fluorodeoxyglucose positron emission tomography findings. Arch. Neurol. 66, https://doi.org/10.1001/archneurol.2009.59.

Price, T., Wee, C.-Y., Gao, W., Shen, D., 2014. Multiple-Network Classification of Childhood Autism using Functional Connectivity Dynamics. Miccai.

Rockwood, K., 2010. Con: can biomarkers be gold standards in Alzheimer's disease? Alzheimers Res. Ther. 2, https://doi.org/10.1186/alzrt40.

Rubinov, M., Sporns, O., 2010. Complex network measures of brain connectivity: uses and interpretations. Neuroimage 52, 1059–1069.

Schaeffer, E.L., Figueiro, M., Gattaz, W.F., 2011. Insights into Alzheimer disease pathogenesis from studies in transgenic animal models. Clinics 66, 45–54.

Seeley, W.W., Crawford, R.K., Zhou, J., Miller, B.L., Greicius, M.D., 2009. Neurodegenerative diseases target large-scale human brain networks. Neuron 62, 42.

Serran-Pozo, A., Frosch, M.P., Masliah, E., Hyman, B.T., 2011. Neuropathological alterations in Alzheimer disease. Cold Spring Harb. Perspect. Med. 1, a006189.

Sporns, O., 2013. Structure and function of complex brain networks. Dialogues Clin. Neurosci. 15, 247–262.

Stam, C.J., 2014. Modern network science of neurological disorders. Nat. Neurosci. 15, 683–695.

Stern, Y., 2006. Cognitive reserve and Alzheimer disease. Alzheimer Dis. Assoc. Disord. 20, https://doi.org/10.1097/00002093-200607001-00010.

Thompson, P.M., et al., 2007. Tracking Alzheimer's disease. Ann. N. Y. Acad. Sci. 1097, 198–214.

Tzourio-Mazoyer, N., et al., 2002. Automated anatomical labeling of activations in SPM using a macroscopic anatomical parcellation of the MNI MRI single-subject brain. Neuroimage 15, 273–289.

Viola, K.L., et al., 2015. Towards non-invasive diagnostic imaging of early-stage Alzheimer's disease. Nat. Nanotechnol. 10, 91–98.

Wee, C.-Y., Yap, P.-T., Zhang, D., Wang, L., Shen, D., 2014. Group-constrained sparse fMRI connectivity modeling for mild cognitive impairment identification. Brain Struct. Funct. 219, 641–656.

Wee, C.-Y., Yap, P.-T., Shen, D., 2016. Diagnosis of autism Spectrum disorders using temporally distinct resting-state functional connectivity networks. CNS Neurosci. Ther. 22, 212–219.

Weimer, D., Sager, M., 2009. Early identification and treatment of Alzheimer's disease: social and fiscal outcomes. Alzheimers Dement. 5, 215–226.

Westman, E., et al., 2011. Multivariate analysis of MRI data for Alzheimer's disease, mild cognitive impairment and healthy controls. Neuroimage 54, 1178–1187.

Wu, G., Kim, M., Wang, Q., Shen, D., 2013. S-HAMMER: hierarchical attribute-guided, symmetric diffeomorphic registration for MR brain images. Hum. Brain Mapp.

Xie, T., He, Y., 2012. Mapping the Alzheimer's brain with connectomics. Front. Psych. 2, 1–14.

Zhang, D., Wang, Y., Zhou, L., Yuan, H., Shen, D., 2011. Multimodal classification of Alzheimer's disease and mild cognitive impairment. Neuroimage 55, 856–867.

Zhu, Y., et al., 2016. Reveal Consistent Spatial-Temporal Patterns From Dynamic Functional Connectivity for Autism Spectrum Disorder Identification. *MICCAI*, Athens.

Zhu, Y., Zhu, X., Kim, M., Wu, G., 2017. A Tensor Statistical Model for Quantifying Dynamic Functional Connectivity. In: Information Processing in Medical Imaing (IPMI). Lecture Notes in Computer Science Book Series (LNCS, Volume 10265), Springer, Boone, NC.

Toward a more integrative cognitive neuroscience of episodic memory

10

Matthew L. Stanley*, Benjamin R. Geib*, Simon W. Davis[†]

Department of Psychology and Neuroscience, Center for Cognitive Neuroscience, Duke University, Durham, NC, United States[*] *Department of Neurology, Center for Cognitive Neuroscience, Duke University, Durham, NC, United States*[†]

CHAPTER OUTLINE

INTRODUCTION

The brain is a complex network, and disparate components of this network dynamically process and integrate information to realize our diverse cognitive capacities and functions. Perhaps one of the most valuable and informative developments in contemporary cognitive neuroscience is the growing shift from univariate analyses for detecting local activations to multivariate network analyses for characterizing the organization of heterogeneous functional interactions. This shift has embraced the complexity of the human brain as an information processing device with both local, specialized processing and more distributed, integrative processing. Multivariate network methods provide a new means to answer questions about the neural underpinnings of cognitive capacities and functions (Medaglia et al., 2015; Sporns, 2014).

Most multivariate functional network analyses of functional magnetic resonance imaging (fMRI) data have attempted to characterize differences in resting-state functional connectivity MRI (rs-fcMRI) patterns between different groups of interest as opposed to functional connectivity patterns during and across cognitive tasks[1]. rs-fcMRI analyses measure correlations in spontaneous low-frequency fluctuations in blood oxygen level dependent (BOLD) signal while individuals are engaged in internally oriented mental activities in the absence of a goal-directed task (Biswal et al., 1995). The observed functional network organization derived from rs-fcMRI data is often deemed "intrinsic", suggesting that the organization emerges in the absence of external stimulation. In contrast, task-based network analyses capture functional connectivity patterns when a person is actively engaged in a cognitive task. There is now considerable evidence that functional connectivity patterns reliably and meaningfully change between resting-state and different cognitive tasks (Bolt et al., 2017; Davis et al., 2017a). Moreover, properties of task-based functional brain networks closely predict behavior and cognitive performance (e.g., Geib et al., 2017a,b; Westphal et al., 2017). Although it has become commonplace for researchers to make inferences about the neural substrates of cognitive processes from rs-fcMRI data, it remains critical to investigate the neural substrates of cognitive processes during cognitive tasks (Davis et al., 2017a). See Chapter 2 by Dale Dagenbach in this volume for a discussion of the use of resting-state and task-based networks for studying the neural bases of cognitive function. In this chapter, we focus our attention on prior studies utilizing task-based fMRI.

One particular domain of cognitive neuroscience that has recently benefited from this shift to a multivariate network approach is episodic memory. Episodic memory allows a person to become consciously aware of earlier personal experiences that occurred at a particular time and place (Tulving, 2002; Moscovitch et al., 2016). Episodic memory can be characterized by three general processes: encoding, consolidation, and retrieval. Episodic encoding refers to the processes whereby events being experienced are converted into information that can be stored for later retrieval, whereas consolidation is the offline process where this information is integrated with an organized lexicon of preexisting memories. Episodic retrieval refers to the processes whereby previously stored events in memory are accessed. Retrieving episodic memories often requires retrieving multiple spatial, contextual, emotional, temporal, sensory, etc. features of the previously consolidated experience. Episodic encoding, consolidation, and retrieval are all complex, multifaceted cognitive pro-

---

[1] In this chapter, we focus primarily on fMRI research for several reasons. First, subcortical structures (e.g., medial temporal lobe regions) are critically important for episodic memory functioning, meaning that fMRI provides a better means of investigating their properties than other methods like electroencephalography (EEG) or magnetoencephalography (MEG). Relatedly, many of the subcortical structures involved in episodic memory functioning are relatively small, so the high spatial resolution of fMRI is better suited for their investigation. Finally, as a more practical consideration, more research has been conducted to investigate the neural substrates of episodic memory using network analyses with fMRI than with EEG or MEG. Nevertheless, we believe that comparisons of network properties during episodic memory tasks between fMRI, EEG, and MEG will prove useful for more completely characterizing how episodic memory functioning is carried out in the brain. EEG and MEG offer much better temporal resolution than fMRI, which might be particularly useful for characterizing fast, transient changes in network organization during some complex, multifaceted memory tasks.

cesses subserved by a widely distributed network encompassing a multitude of cortical and subcortical brain regions (Tulving, 2002; Geib et al., 2017b; Moscovitch et al., 2016)[2].

Network neuroscience and the formalisms of graph theory offer a theoretical framework and an extensive set of tools to make new discoveries about how the brain carries out episodic memory. These tools include a diverse repertoire of metrics for characterizing both local, segregated functioning between some small subset of brain regions and more global integrative functioning throughout the entire brain (Sporns, 2011; see Rubinov and Sporns, 2010 for a comprehensive discussion of these graph metrics). Different graph metrics can provide unique information about the properties of individual nodes (i.e., brain regions), subsets of nodes in a network, and the entirety of the network. However, by adopting this framework and these tools, there is a fundamental shift in how we approach task-based neuroimaging data. With a growing number of researchers utilizing graph theoretic measures to investigate how the brain subserves episodic memory (and cognition more generally), it is now more important than ever to systematically consider what has been offered by the predominant network analysis methods currently in the field, and how these techniques complement and conflict with earlier methods of examining neural correlates of memory functioning.

Here, we discuss some underlying assumptions and motivations across three general methodologies in the cognitive neuroscience of episodic memory: (1) univariate, voxel-wise analyses; (2) direct connections between pairs of voxels/regions (i.e., bivariate connectivity analyses and seed-based analyses); and (3) multivariate network metrics used to characterize complex patterns of interactions between many voxels/regions. The introduction of each method in the recent history of cognitive neuroscience has purportedly provided a shift in the science. We believe that there has been relatively little discussion regarding the differences and similarities in what each methodology offers—conceptually and theoretically—for characterizing how episodic memory occurs, and that this explicit discussion is necessary to harmonize the wealth of knowledge gained from univariate analyses of episodic memory with the dynamic, complex future afforded by network neuroscience. Fig. 1 broadly characterizes major findings, strengths, and weaknesses of these three approaches. We begin this effort by first delineating what each methodology offers, and then highlighting their respective advantages and disadvantages. Next, we argue that, by adopting an informed multivariate network approach to task-related connectivity information, new and unique insights can be obtained into the neural underpinnings of episodic memory. Finally, we use the progression of the summarized research to point toward a future for the cognitive network neuroscience of episodic memory that emphasizes distributed, integrative information processing in addition to more segregated, localized information processing.

---

[2] In this chapter, we focus on episodic encoding and retrieval, because they have been directly investigated in the literature using all methods we survey (e.g., univariate activation analyses, bivariate connectivity, and multivariate network analyses).

	Activation models of episodic memory	Seed-based connectivity models of episodic memory	Network models of episodic memory
Key findings specific to each modality	• Hippocampus is involved in relational memory, pattern completion (Eichenbaum et al., 1994; Bakker et al., 2008) • Perirhinal, lateral temporal cortex involved in mediating familiarity-based recognition (Staresina et al., 2012) • Ventrolateral PFC controls access to semantic representations (Badre and Wagner, 2007)	• Frontoparietal connectivity predicts recollection accuracy (King et al., 2015) • Distinct connectivity profiles for subregions in the MTL (Libby et al., 2012) • Compensatory increases in frontotempoal connectivity during encoding in aging (Dennis et al., 2008) • Hippocampal-VPC connectivity underlying bottom-up attention for memory (Cabeza et al., 2008)	• Hippocampus becomes better integrated with the rest of the network to facilitate successful recognition memory (Geib et al., 2017b) • Frequency-specific network interactions during episodic retrieval (Watrous et al., 2013) • A more globally integrated network architecture facilitates successful memory retrieval (Geib et al., 2017a; Westphal et al., 2017)
Strengths	• Localizes brain function with mm precision • Identifies concurrent activations in disparate regions • More sophisticated techniques (MVPA, RSA) can identify information content in localized patterns of activations • Offers a concrete interpretation of BOLD-related responses to mnemonic information	• Identifies consistent co-activation in disparate regions • Identifies connections between regions at a voxel-wise level • Possible to examine across-trial modulation of global regressors (e.g., vivideness, memory strength) via PPI or single-trial regression	• Considers all possible connections simultaneously • Identifies behaviorally meaningful assembly behavior • Accounts for indirect connections • Can account for the potential influence of all brain regions in a single metric
Weaknesses	• Ignorant of hierarchical temporal relationships • Integrative functioning is key to episodic memory functioning but cannot be addressed by activation models	• Bivariate relationships between two regions may be mediated by a third unidentified region • Fails to isolate consistent motifs or families of connected regions	• Difficult to isolate trial-specific responses • Often rely on large ROIs comprising many individual voxel units, thereby losing information • Current over-reliance on resting state graphs preclude meaningful conclusions on task-based responses

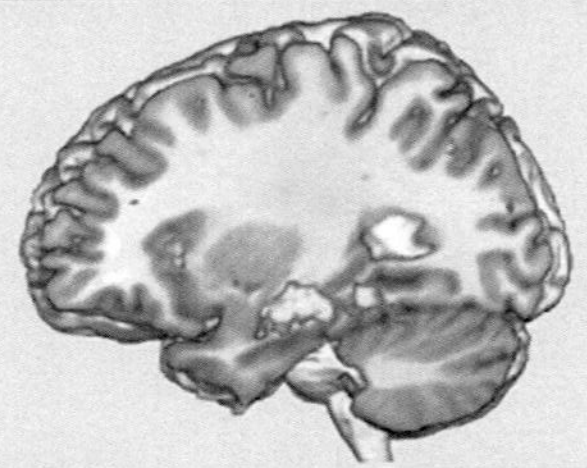

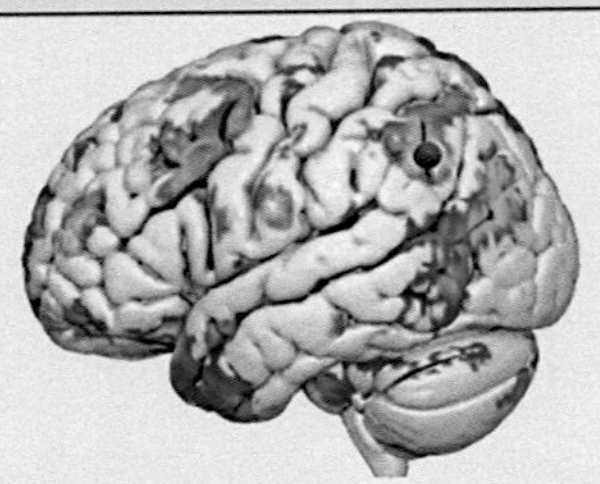

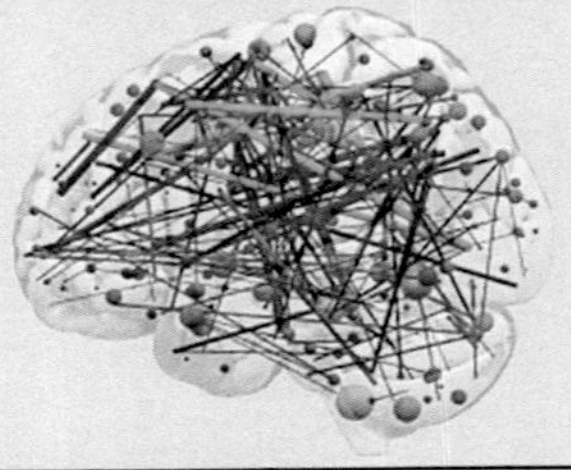

FIG. 1

Summary of key findings and qualities supporting univariate activation, seed-based connectivity, and network-based approaches to the study of episodic memory function.

UNIVARIATE ACTIVATION ANALYSES

Research in the cognitive neurosciences has traditionally operated under the assumption that cognitive processes and subprocesses can be mapped onto discrete, localized neural structures (Anderson, 2007, 2015; Bergeron, 2007). In other words, the transformations of inputs to outputs defining a particular process occur in a localized brain region (Anderson, 2007). This overarching assumption guiding the science is commonly called *localizationism*. Generally speaking, univariate activation analyses have been implemented to localize specific psychological processes and subprocesses to specific neural structures. By localizing all processes and subproceses to specific brain regions, researchers would, in theory, be able to decompose the entire system (i.e., the brain) into parts; characterize those parts in functional, psychological terms; and then explain the behavior of the system by appealing to the functions of the parts (Burnston, 2016). Such a research program provides a clear path for moving the science forward, all while producing falsifiable predictions to allow researchers to continuously refine, improve, and perfect region-specific theories of cognitive processes[3].

To better understand the basis for making claims about activations and the localization of cognitive processes to neural structures, it will be useful to briefly review how inferences about neural functioning are made using fMRI. Most models of task related activity are estimated by the BOLD contrast, which closely correlates with extracellular local field potentials (LFPs) and is used to make inferences about the underlying but unobserved neuronal activations (Ojemann et al., 2013). The hemodynamic response function (HRF) and variants thereof (gamma functions, balloon models, and so on) are used as a basic means of translating the intensity of the signal measured with BOLD fMRI into a workable model of the underlying assumptions about oxygen consumption that reflect neural activity (Worsley and Friston, 1995). The HRF serves as the basis for claims about activations in the overwhelming majority of univariate activation analyses in the cognitive neuroscience of episodic memory. The shape of the HRF is relatively fixed; only the height (i.e., amplitude) of the response is allowed to vary (e.g., in parametric models of activity). Some neuroimaging studies of episodic memory have taken more flexible approaches because of underlying assumptions about the time courses of the cognitive processes under study. For example, finite-impulse response (FIR) models of functional activity make

---

[3] Some advocates of localizationism in cognitive neuroscience have assumed that, once requisite granularity is achieved, each circumscribed brain region will be responsible for carrying out a single, univocal process (i.e., a one-to-one mapping of cognitive processes to neural structures). Under weaker views of localizationism, however, it is only necessary that at least some brain regions are responsible for carrying out some subset of cognitive processes. Furthermore, proponents of weaker views of localizationism can readily grant that other fundamental organizational principles beyond strict neural localization bring about certain cognitive processes or behaviors. Although explanation in the cognitive neurosciences has traditionally revolved around ascribing functions or processes to decomposed, localized neural structures, the weak localizationist could grant that other varieties of explanation will better account for some observations and regularities.

no assumptions about the shape or temporal distribution of the hemodynamic response. Model choice is typically driven by the research questions.

What exactly do cognitive neuroscientists learn about the neural substrates of episodic memory by performing univariate activation analyses? Traditionally, the cognitive neuroscience of episodic memory has progressed by characterizing the processes or functions carried out by specific, localized brain regions using psychological manipulations. For example, using fMRI a researcher might design a study in which participants engage in one cognitive task, A, that involves certain cognitive processes, $X_1 \ldots X_n$. Participants would then engage in another comparable task, B, that is assumed to involve all the processes involved in A plus another process of interest, Y. By subtracting the brain activity of A from B, the resultant neural activations are thought to be responsible for carrying out process Y. This strategy, which is commonly referred to as the "subtraction method," allows researchers to find either single or double dissociations to identify particular brain regions thought to carry out specific cognitive processes (Klein, 2010). This approach provides no explicit information about how different brain regions functionally cooperate or interact to carry out cognitive processes. Instead, this approach only allows for the identification of some isolated set of neural activations.

Using this subtraction method, dissociations have been identified between different processes, such as episodic encoding versus retrieval (Eldridge et al., 2005), recall versus recognition (Cabeza et al., 2003), object versus spatial location encoding (Buffalo et al., 2006), and word versus pattern encoding (Iidaka et al., 2000), among others. Moreover, with these dissociations serving as evidence, specific processes have been ascribed to localized brain regions, such as the hippocampus in processing relational memory representations (Eichenbaum et al., 1994; Bakker et al., 2008), perirhinal cortex in integrating perceptual features in memory (Bussy et al., 2005) and familiarity-based recognition decisions (Staresina et al., 2013a, b), the ventrolateral prefrontal cortex in controlling access to semantic representations (Badre and Wagner, 2007), and the ventral parietal cortex in bottom-up attention during memory retrieval (Cabeza et al., 2012). Because episodic encoding and retrieval are such complex cognitions, the cognitive neuroimaging literature has identified a multitude of broadly distributed brain regions—spanning all lobes of the brain—thought to carry out component processes and subprocesses. Across these studies, the underlying assumption is that complex behaviors can, in fact, be broken down into component processes, and that these processes can be directly and straightforwardly mapped onto specific neural structures. Accordingly, theories about specific brain regions are developed and refined, often without reference to how those specific brain regions are situated within a larger network comprised of many other disparate but functionally cooperating components.

This localization strategy has worked well when a given brain region is thought to perform some computation or transformation, and when reference to how that brain region is embedded within the network is not immediately necessary for specifying the cognitive function. For example, inferior temporal cortex is thought to carry out object recognition and classification (Ishai et al., 1999; Kornblith and Tsao, 2017;

Majaj et al., 2015), and characterizing how inferior temporal cortex influences, modulates, acts on, excites, inhibits, etc. other brain regions is not immediately necessary for specifying the neural instantiation of this particular cognitive function. Localizing a cognitive function to a brain region in this way provides for a relatively straightforward mechanistic explanation for how the brain carries out object recognition and classification in the service of some more general task. Certainly, other brain regions transmit information to the inferior temporal cortex so that it can carry out its function, and processed information within the inferior temporal cortex is sent to other brain to carry out the more general task. But rigorously characterizing the connectivity properties of the inferior temporal cortex is not necessary for claiming that a function of the inferior temporal cortex is to carry out object recognition and classification. Characterizing the connectivity properties of the inferior temporal cortex will, however, allow for a more complete understanding of how the more general task is achieved (Shimamura, 2010). Moreover, the localization of cognitive functions can be carried out using relatively well-established and straightforward statistics, making such work easily accessible and computationally inexpensive.

Localization of psychological processes to specific neural structures provides valuable information when it can be found, but it is not the only way to investigate, explain, or predict how the brain carries out cognition. By localizing processes to specific neural structures and developing region-specific theories, segregation has taken precedent over functional cooperation, or *integration*—the sharing or transfer of information between disparate brain regions. The ways in which brain regions share and transfer information plays an important role in episodic memory, and the joint contributions of segregated and integrative functioning are thought to be necessary for achieving cognition more generally (Sporns, 2011, 2014). The statistical methods implemented in univariate activation analyses (and some multivariate techniques such as representational similarity analysis, multivariate pattern analysis, and partial least squares) preclude researchers from making claims about functional connections between brain regions. Instead, these methods are employed only to identify transient increases and decreases in activity within specific brain regions.

BIVARIATE AND SEED-BASED FUNCTIONAL CONNECTIVITY ANALYSES

Episodic memory does not only depend on activations within a discrete set of brain regions to which certain processes have been localized. The ability for those brain regions and others to share information and to influence each other has a critical role in episodic memory (Geib et al., 2017a; Westphal et al., 2017). Functional connectivity between brain regions has been broadly defined as a temporal relationship, or statistical dependency, between spatially remote neurophysiological events (Friston, 2011). This relationship is typically measured based upon coactivation; however, a related construct, effective connectivity, captures the causal directionality of information transfer between brain regions (for a directed connection; Friston, 2011). In general, functional connectivity analyses characterize how and to what extent brain regions share information and influence each other.

Functional connections between brain regions can be conceptually characterized in two ways: (1) whether a connection exists at all between the brain regions (for a binary, unweighted connection) and (2) the extent to which the brain regions are connected (for a weighted connection). As a point of contrast, a structural connection might represent the number of white matter tracts between two brain regions, the average fractional anisotropy of tracts connecting two brain regions, or other similarities in structural neural properties between two brain regions (Davis et al., 2009). Univariate activation analyses can reveal whether two distinct brain regions are active as separate, discrete entities during a task (relative to some other condition or control), but such analyses cannot characterize whether or to what extent those two brain regions interact and integrate information. In fact, even if neither of two brain regions are considered active after performing a subtraction, those two regions could still have a strong functional or effective connection between them. And that functional or effective connection could be integral in carrying out the cognitive task.

The notion of functional connectivity between cortical regions is almost as old as some of the first event-related studies in the cognitive neuroscience of memory (Friston et al., 1995, 1997). However, the widespread use of a robust, trial-wise measure of functional connectivity was slow to emerge. Here, we briefly describe the three most widely used methods for identifying functional connections, all of which initially served to inform questions about the relationships between a priori selected pairs of regions based upon activation analyses.

First, in block designs, which are among the oldest functional imaging paradigms, statistical dependencies in the fMRI time series between brain regions can be computed within an entire run or block, often alternating between different task and rest (or baseline) intervals. Statistical dependencies in the time course of signal between brain regions, computed over the duration of the block, are commonly estimated using Pearson's correlation, mutual information, or other metrics. The disadvantage of this technique for episodic memory studies is that it is difficult (if not impossible) to separate successful from unsuccessful memory on a trial-wise basis. Second, psychophysiological interaction (PPI) is a conceptually related technique (Friston et al., 1997) used to modulate the fMRI time series with a regressor representing task events convolved with a standard HRF. PPI allows for the modeling of positive (remembered) or negative (forgotten) modulations of the fMRI time series. This method has since been implemented to identify large sets of functional connections for whole-brain multivariate network analyses (Davis et al., 2017b). Third, the beta-series method (or "single-trial modeling"; Rissman et al., 2004) is used to analyze event-related fMRI data during cognitive tasks by isolating individual trials within a fMRI run and typically convolving those single trials with a standard hemodynamic response function (HRF). Individual beta-values reflect the fit shape of the hemodynamic response evoked by a given trial, and correlations in each beta-series calculated between two brain regions indicate the presence of a functional connection. As with PPI, the beta-series method has been implemented to identify large sets of functional connections for whole-brain multivariate network analyses (Geib et al., 2017a,b). Unfortunately, a rigorous comparison of these techniques in the episodic memory domain is currently absent.

The simplest functional connectivity analyses—regardless of the particular method used to identify functional connections—characterize the relationship between only two a priori identified brain regions that form a functional partnership to carry out some more general process (i.e., pair-wise relationships). These analyses tend to be hypothesis-driven, and the two brain regions are typically a priori selected from the outcome of a univariate activation analysis. As such, the cognitive interpretation of the pair-wise relationship is fundamentally dependent upon existing region-specific theories about localized cognitive processes. Each region is thought to perform its own complementary subprocess in service of the more general process, but the interaction between the regions is necessary to facilitate the more general process (Moscovitch et al., 2016). In this way, the neural instantiation of cognition is fundamentally linear, additive, and hierarchical. Although more general, complex processes are built up from less complex subprocesses, the most basic, fundamental level of the hierarchy is characterized by the localization of specific subprocesses to neural structures.

These pair-wise functional partnerships are thought to rapidly assemble when needed and then disassemble when no longer needed (Cabeza and Moscovitch, 2013). Furthermore, a given brain region A might dynamically form a functional partnership with another brain region B to carry out some general process X. After process X is complete and the functional partnership between A and B has disassembled, A might form a new functional partnership with another brain region C to carry out some new process Y. In this way, a given brain region might have many different functional partnerships that dynamically assemble and disassemble in accordance with the demands of the immediate cognitive task. However, when brain region A dynamically and transiently forms these different functional partnerships with brain regions B and C, it is generally assumed that these functional partnerships support related processes during the different cognitive tasks (Cabeza and Moscovitch, 2013). The specific relationship between those processes can typically be described in cognitive terms.

Many pair-wise functional partnerships have been characterized in the episodic memory literature. For example, some have suggested that a functional partnership is formed between the ventrolateral prefrontal cortex and the hippocampus to carry out the encoding and retrieval of new episodic information (Dennis et al., 2008, Rugg and Vilberg, 2013). During this process, the ventrolateral prefrontal cortex is assumed to carry out an elaboration process that facilitates the storage of information in the hippocampus (Cabeza and Moscovitch, 2013). Both brain regions carry out their own subprocesses, but together they carry out a more complex process via their interaction. The strength of the functional connection between the ventrolateral prefrontal cortex and the hippocampus during episodic encoding even predicts memory performance in a later recognition test (Grady et al., 2003). As another example, to facilitate episodic memory retrieval, the hippocampus and angular gyrus form a functional partnership. The hippocampus is thought to be responsible for recovering episodic memory details, whereas the angular gyrus is thought to operate on the re-covered memory details through bottom-up attention (Cabeza et al., 2008; although see Vilburg and Rugg, 2008 as well as Simons et al., 2010 for alternative accounts). There is still debate over the exact process localized to angular gyrus; but regardless, the angular gyrus is thought to act on the hippocampus output to facilitate episodic memory retrieval. Seed-based connectivity methods have also been instrumental in dissociating the connectivity profiles of anterior versus posterior MTL regions (Libby et al., 2012). Certainly, researchers are not limited to characterizing processes carried out by just two brain regions. Hypothesis-driven investigations have char-acterized the extent to which more than two brain regions functionally or causally interact to carry out more complex processes in episodic memory (e.g., Leal et al., 2017; Staresina et al., 2012, 2013a, b).

Beyond characterizing only a few functional connections between a few a priori selected brain regions, cognitive neuroscientists have also used seed-based analyses to quantify the extent to which one a priori selected brain region (i.e., the seed-region) is directly functionally connected with *all* other brain regions (e.g., Bai et al., 2009; Sneve et al., 2015). The brain regions that exhibit strong functional connec-tions to the seed-region are commonly displayed in a functional connectivity map. Although seed-based analyses are typically more exploratory given that they char-acterize the extent to which one brain region is directly functionally connected with all other brain regions, the seed-region must be a priori selected. In practice, cogni-tive neuroscientists typically select a seed-region based upon a univariate activation analysis. Accordingly, region-specific theories about the cognitive process localized to the seed-region drive both the justification for the seed-based analysis and the interpretation of results from the seed-based analysis.

For example, King et al. (2015) completed an extensive series of seed-based functional connectivity analyses across three experiments to characterize those di-rect functional connections involved in recollection. Their seed-regions were a priori selected based on the results of univariate activation analyses from prior research (Spaniol et al., 2009; Rugg and Vilberg, 2013). That prior research had suggested that those specific regions were responsible for carrying out successful recollection. In other words, the seed-regions selected were thought to be *the* regions performing the cognitive processes necessary for successful recollection. Seed-regions included left angular gyrus, medial prefrontal cortex, hippocampus, middle temporal gyrus, and posterior cingulate cortex. Although their results showed that the set of seed-regions displayed recollection-related increases in functional connectivity with each other, the seed-regions also displayed recollection-related increases in functional connec-tivity with several other unexpected brain regions, including dorsolateral prefrontal cortex, dorsolateral posterior cingulate cortex, dorsal anterior cingulate cortex, and extrastriate visual cortex. Thus, the results from these more exploratory seed-based analyses offered new information about the how successful recollection is subserved in the brain; this valuable information would not have been obtained if the researchers had just examined pair-wise functional connections exclusively between a constrained set of a priori selected brain regions.

Furthermore, comparing their own univariate activation results (i.e., not brain regions based on the preexistent meta-analyses) to their functional connectivity results, King et al. (2015) found that those brain regions displaying recollection-related univariate activation effects were typically *not* the same brain regions displaying recollection-related connectivity effects. So, within their own datasets, more brain regions and different brain regions came out of the recollection-related connectivity analyses relative to the univariate activation analyses. This suggests that functional connectivity analyses should not be entirely constrained by or derived from assumptions about which specific processes are localized to particular neural structures. Integrative processing between some set of brain regions may be necessary for facilitating successful recollection, even if those same brain regions do not exhibit univariate activation effects.

In a complementary study, Sneve et al. (2015) scanned participants during an associative encoding task and tested memory for those associations either hours or weeks later. Although localized univariate activations in disparate brain regions were associated with memory success after the brief delay, no univariate activations were associated with memory success after the long delay. In contrast, using a seed-based analysis, strong functional connections between the hippocampus and disparate occipital, parietal, and temporal regions were associated with successful recollection after the long delay. These functional connectivity results were interpreted as being responsible for establishing which of these memory traces will take a more permanent form and remain accessible in the long term. As in King et al. (2015), these results suggest that functional connectivity analyses provide new, valuable information about how the brain facilitates episodic memory distinct from the univariate activation results.

Regions that are functionally connected to facilitate successful recollection are often not the same regions that show univariate activation effects. Furthermore, it's possible (and likely) that a seed-based analysis performed on *any* given brain region during any cognitive task would yield significant functional or effective connections related to that cognitive task—regardless of whether or not the region exhibits a univariate activation effect. This makes intuitive sense if a researcher grants that the function of many brain regions is to influence, interact with, drive, modulate, control, act on, inhibit, suppress, excite, etc. at least one other brain region in service of the more general task at hand. If the function of a region in carrying out a cognitive task is explained by the ways in which it interacts with other regions, then identifying functional or effective connections between that region and other regions is sufficient evidence for asserting that it carries out the function. Identifying a localized activation via a subtraction without reference to how that brain region dynamically interacts with other brain regions in the network will not provide the evidence needed to ascribe the function to the region.

For example, if King et al. (2015) had also used the dorsolateral prefrontal cortex as a seed-region, they might have discovered that this additional seed-region is functionally connected with other disparate brain regions above and beyond those already identified. As such, it is possible that existing results from seed-based analyses only depict a small piece of the larger picture. This suggests that many more functional connections might be involved in carrying out cognitive tasks than has traditionally been assumed (Geib et al., 2017b). Although seed-based analyses have provided valuable new insights into how the brain facilitates episodic encoding and retrieval, cognitive neuroscientists applying the method have typically been concerned only with characterizing changes in direct, immediate functional connections emanating from one or a few seed-regions. While direct connections represent the immediate functional partners of a brain, two brain regions might only be indirectly connected through a third brain region (Paolini et al., 2014). Generally speaking, brain regions carrying out some function may require information to be transferred from brain region A to brain region C but only after that information is partially processed or transformed at brain region B (Sporns, 2011; Stanley et al., 2013). It's even possible that information processing and transformation occurs at multiple different intermediate brain regions before going from brain region A to brain region C. To provide a simplistic example, Geib et al. (2017b) showed that, although a spatially disparate set of brain regions were directly functionally connected to the hippocampus to carry out successful episodic memory retrieval, many regions that were directly functionally connected to the hippocampus were not functionally connected to each other. For example, although medial prefrontal cortex and inferior temporal cortex showed increases in direct functional connectivity with the hippocampus to facilitate successful memory retrieval, the medial prefrontal cortex and the inferior temporal cortex were not themselves directly functionally connected with each other. Nevertheless, it is plausible that the medial prefrontal cortex and inferior temporal cortex had some indirect effect on each other through the hippocampus. This was not a special or atypical case. Most of the brain regions exhibiting significant functional connections with the hippocampus to facilitate successful memory retrieval were not themselves directly connected to each other (Geib et al., 2017b).

Because there are physical, structural limits on the possible set of direct functional or effective connections that could exist in a brain network, it is not possible for any given brain region to directly share or transfer information with just any other brain region. Even if a functional or effective connection is identified between brain regions X and Z, it is possible that those regions can only interact through another brain region Y due to anatomical constraints. Recent research using noninvasive imaging technologies has shown that whole-brain white-matter networks in humans are exceedingly sparse (Sporns, 2011; Davis et al., 2017a, b). Brains are efficient, economical structures organized to minimize wiring costs while supporting the capacity for high dynamic complexity (Bassett and Bullmore, 2006; Bullmore and Sporns, 2012; Sporns, 2011). Yet some brain regions that are not directly connected by underlying structure likely share and transfer information in an indirect way to carry out more complex cognitions, such as episodic encoding and retrieval (Geib et al., 2017a,b). Some information about indirect connections could, in theory, be acquired by performing a series of seed-based analyses on the same data set. However, this approach will likely result in a rapid proliferation in identified (significant) functional connections, as more and more seed-based analyses are performed on different seed-regions. Neatly describing and summarizing so many direct and indirect functional connections between such diverse, spatially disparate brain regions would be a challenging endeavor. Fortunately, graph theory provides a rigorous set of tools for characterizing how large sets of functional connections are organized topologically. That is, graph theoretic metrics can provide information about how each brain region is directly and/or indirectly connected with all other brain regions.

MULTIVARIATE NETWORK ANALYSES

Several task-based functional neuroimaging studies have identified changes in the strength and direction of specific functional connections between one or more a priori selected seed-regions and other disparate brain regions that facilitate episodic memory functioning. These studies have provided new insights into how the brain enables episodic memory to happen, above and beyond what has been learned from univariate activation effects. In investigating the integrative properties of the brain that play a role in episodic memory, we believe there is additional value in adopting multivariate network methods to make new discoveries. Large-scale complex network analyses embracing the formalisms of graph theory are shedding new light on the neural instantiation of episodic memory. This allows for a more exploratory, whole-brain approach in characterizing how different patterns of functional interactions enable episodic memory, because researchers can take into account *all* possible connections between all brain regions simultaneously. In contrast, with seed-based analyses on some subset of brain regions, it is possible that significant connections between uninvestigated brain regions will be left undiscovered, even though those connections could be highly predictive of cognitive performance or behavior. As such, this approach will provide an unnecessarily restricted and incomplete picture of the connectivity properties associated with cognitive performance or behavior. If a significant functional connection exists between any two brain regions, it can be incorporated into multivariate network analyses in a meaningful way. It is not necessary to a priori select some subset of brain regions to complete a meaningful analysis of neuroimaging data.

All graphs that model real-world networks are comprised of differentiable elements of the system (nodes) and pairwise relationships between those elements (edges). Adjacency matrices, or matrices in which both rows and columns are labeled by an ordered list of elements (nodes), represent the full set of nodes and connections in a network (Bullmore and Sporns, 2009; Pessoa, 2014). In task-based functional brain networks, nodes (i.e., brain regions) must be delineated and edges (i.e., functional or effective connections) must be measured prior to characterizing the topological properties (i.e., the ways in which edges in the network are organized) of the system using graph theoretic measures (Pessoa, 2014; Sporns, 2011). Graph theory provides a comprehensive set of tools to quantitatively characterize the topological properties of complex systems such as the brain (Rubinov and Sporns,

2010; Telesford et al., 2011). In practice, graph theoretic network analyses have been conducted by incorporating all of the brain space with a detectable signal and by incorporating only some subset of a priori identified regions (Stanley et al., 2013).

A complex functional brain network is a massively nonlinear system. In this system, a reorganization of one brain region's direct functional connections will have a reverberating effect throughout the system, likely bringing about changes in the system's local and global topological properties. This could have an effect on whether and how information is shared between disparate brain regions within the larger system—even for brain regions not directly functionally connected to the brain region that reorganizes its direct connections. In other words, the topological properties ascribed to each brain region are dependent upon many other brain regions and their connections. No brain region operates as a completely isolated entity unless it has exactly zero functional connections in the network. Because of nonlinearity and dependency in complex brain networks and because all functional connections between all brain regions filter into the calculations of some graph metrics, indirect connections have a significant role to play in determining the topological properties of the systems. This represents a unique benefit of the multivariate network approach. Diverse, spatially disparate brain regions may functionally cooperate to carry out some cognition, but that cooperation may depend on indirect connections through other nodes in the network.

Adopting multivariate methods from network science does not preclude the cognitive neuroscientist from developing region-specific theories. Cognitive neuroscientists have characterized how certain topological properties of a specific brain region change and reorganize within the larger network for some specific purpose. As such, region-specific theories can still be developed within a multivariate network framework, but those theories will involve references to how the region is situated within the larger network (i.e., how the region directly and indirectly interacts with other brain regions in the network). For example, the hippocampus is thought to play a pivotal role in successful episodic memory retrieval, but that role demands that the hippocampus functionally interact with disparate other brain regions for diverse purposes. Geib et al. (2017a) showed that, to facilitate the retrieval of vivid relative to dim visual images from memory, the hippocampus drastically reorganized its set of functional connections to (1) exhibit a greater capacity for efficient information integration with disparate other brain regions and (2) become a more convergent structure for information integration. Importantly, no graph theoretic measures were correlated with univariate activation effects, and there was no discernable difference in univariate activation for the hippocampus between dimly and vividly remembered items. Similarly, in another study, Geib et al. (2017b) showed that successfully remembering relative to forgetting words was associated with significant changes in the connectivity profile of the hippocampus and more efficient communication between the hippocampus and the rest of the network. As in Geib et al. (2017a), this was the case even though no graph theoretic measures were correlated with univariate activation effects, and there was no discernable difference in univariate activation for the hippocampus between remembered and forgotten items. In both studies conducted by Geib and colleagues, a specific function was still ascribed to a specific brain structure (i.e., the hippocampus), but the function was cached out in terms of the integrative role of the region within the larger network. In other words, the function of the brain region in carrying out the cognitive task is to influence, interact with, drive, modulate, control, act on, inhibit, suppress, excite, etc. other regions for some purpose.

Graph metrics characterizing global topological properties of entire functional brain networks also change in reliable ways that track task performance (Geib et al., 2017a,b; Schedlbauer et al., 2014; Westphal et al., 2017; Meunier et al., 2014). More specifically, an overall quantity and strength of connections in the entire functional brain network was shown to predict successful memory for spatiotemporal contextual detail (Schedlbauer et al., 2014), the vivid retrieval of pictures from memory (Geib et al., 2017a), and the successful retrieval of concrete words (Geib et al., 2017b). Complementary findings have suggested that more global patterns of oscillatory coupling between many disparate brain regions facilitate the retrieval of episodic memories (Watrouos and Ekstrom, 2014; Watrous et al., 2013). A more globally integrated functional network architecture with less distinctive segregated communities has been associated with the successful retrieval of concrete words (Geib et al., 2017b), successful odor recognition memory (Meunier et al., 2014), the successful retrieval of words and associated imagery context (Westphal et al., 2017), and the retrieval of subjectively vivid visual images of scenes (Geib et al., 2017a). In these more globally integrated functional brain networks, a brain region can share and transfer information with any other brain region in relatively few steps, and brain regions are less inclined to form small groupings that only share information with each other. Broadly distributed information processing facilitates episodic memory retrieval.

Across these diverse episodic retrieval tasks, there is clear and consistent evidence that a more globally integrated network architecture is associated with better memory performance. This might suggest that a domain-general neural property necessary for episodic memory retrieval is a more globally integrated functional network architecture. To be clear, this interpretation does not entail that this is the singular and sufficient explanation for how the brain makes episodic memory retrieval happen. Even if a global shift in a topological property is detected that predicts memory performance, that change may be predominantly driven by a reorganization of a specific subset of connections in the network or even the connectivity profile of a single network node. The ability to comprehensively explain how a functional brain network is related to episodic encoding and retrieval functioning will likely require appeals to changes in specific connections, regional shifts in topological properties, and more global shifts in topological properties (Geib et al., 2017b). Furthermore, better characterizing what drives global shifts in topological properties will improve our understanding of neural substrates of episodic memory functioning.

CONCLUSIONS AND FUTURE DIRECTIONS

We have attempted to broadly and conceptually characterize three general methodologies in the cognitive neuroscience of episodic memory: (1) univariate, voxel-wise analyses; (2) direct functional connectivity analyses between specific pairs of voxels/regions; and (3) multivariate graph theoretic analyses used to characterize complex patterns of interactions between many voxels/regions. The development and implementation of each methodology in the recent history of cognitive neuroscience has offered new and unique insights about how the brain carries out episodic encoding and retrieval. The cognitive neuroscience of episodic memory has traditionally focused on localizing cognitive processes and subprocesses to localized activations without reference to how those brain regions interact—both directly and indirectly—with disparate other brain regions. The advent of new, multivariate methods in network neuroscience are allowing researchers to gain novel insights into how information is shared and integrated between disparate brain regions to make episodic memory happen. Graph theory has provided a comprehensive mathematical framework for understanding how heterogeneous interactions within the brain are organized and how network architectures change across different task states and conditions.

To maximize the potential of graph theoretic network analyses for characterizing the neural substrates of episodic memory functioning, there are still significant hurdles to overcome. More general issues that face the field involve adequately defining nodes and edges for subsequent network analysis (Pessoa, 2014; Smith, 2012; Stanley et al., 2013). In practice, how researchers define nodes and edges for network analysis typically depends on the research question, but the proliferation of different strategies for defining nodes and edges makes it challenging to straightforwardly compare topological properties between different studies. Furthermore, because brain networks are nonlinear systems, the topological properties of brain networks will likely differ as a function of how much brain space is incorporated into the network analysis. For example, although some researchers have only used a relatively small portion of brain space to define nodes and then investigate network topology during episodic memory retrieval (Schedlbauer et al., 2014), others have incorporated the entire cortex and many subcortical regions into network analyses during episodic memory retrieval (Geib et al., 2017a,b). This further renders the comparison of topological properties between different studies problematic. Moreover, because the cognitive neuroscience of episodic memory has progressed largely by localizing functions to specific brain regions using activation analyses, it has been challenging to relate those region-specific activation results to whole-brain topological properties. Although some suggest that characterizing the topology of brain networks offers a totally new kind of explanation for the neural instantiation of cognitive phenomena that should be independent of explanations offered from activation analyses (Huneman, 2010; Rathkopf, 2015), others argue that appeals to topological properties offer a straightforward extension of mechanistic explanations based upon activation results (Craver, 2016). How localized activations and whole-brain topology are related to each other (if at all) remains a matter of debate.

Network neuroscience and the formalisms of graph theory offer both a theoretical framework and an extensive set of tools to make new discoveries about how the brain carries out episodic memory (Rubinov and Sporns, 2010). This theoretical framework and set of tools will allow cognitive neuroscientists to better understand how episodic memory emerges in development, how episodic memory happens in healthy adults, and how episodic memory declines with both normal and pathological aging. Although we focus on fMRI research in this chapter, the same graph metrics can be implemented for brain networks derived from EEG and MEG data (Bullmore and Sporns, 2009; Sporns, 2014). Assuming adequate computational power, graph metrics can be computed on networks at any spatial or temporal scale, and different kinds of information can be extracted from these different networks. Comparisons in topological properties between brain networks derived from fMRI, EEG, and MEG will provide a more complete picture of how the brain carries out episodic memory functioning. There are still significant hurdles to overcome, but multivariate network models have a lot to offer the cognitive neuroscience of episodic memory.

REFERENCES

Anderson, M.L., 2007. The massive redeployment hypothesis and the functional topography of the brain. Philos. Psychol. 202, 143–174.

Anderson, M.L., 2015. After Phrenology: Neural Reuse and the Interactive Brain. MIT Press, Cambridge, MA.

Bakker, A., Kirwan, C.B., Miller, M.I., Stark, C.E., 2008. Pattern separation in the human hippocampal CA3 and dentate gyrus. Science 319, 1640–1642.

Badre, D., Wagner, A.D., 2007. Left ventrolateral prefrontal cortex and the cognitive control of memory. Neuropsychologia 45, 2883–2901.

Bai, F., Zhang, Z., Watson, D.R., Yu, H., Shi, Y., Yuan, Y., Zang, Y., Zhu, C., Qian, Y., 2009. Anormal functional connectivity of hippocampus during episodic memory retrieval processing network in amnestic mild cognitive impairment. Biol. Psychiatry 65, 951–958.

Bassett, D.S., Bullmore, E., 2006. Small-world brain networks. Neuroscientist 12, 512–523.

Bergeron, V., 2007. Anatomical and functional modularity in cognitive science: shifting the focus. Philos. Psychol. 20, 175–195.

Biswal, B., Yetkin, F.Z., Haughton, V.M., Hyde, J.S., 1995. Functional connectivity in the motor cortex of resting human brain using echo-planar MRI. Magn. Reson. Med. 34, 537–541.

Bolt, T., Nomi, J.S., Rubinov, M., Uddin, L.Q., 2017. Correspondence between evoked and intrinsic functional brain network configurations. Hum. Brain Mapp. 38, 1992–2007.

Buffalo, E.A., Bellgowan, P.S., Martin, A., 2006. Distinct roles for medial temporal lobe structures for objects and their locations. Learn. Mem. 131, 638–643.

Bullmore, E.T., Sporns, O., 2009. Complex brain networks: graph theoretical analysis of structural and functional systems. Nat. Rev. Neurosci. 10, 186–198.

Bullmore, E.T., Sporns, 2012. The economy of brain network organization. Nat. Rev. Neurosci. 13, 336–349.

Burnston, D.C., 2016. A contextualist approach to functional localization in the brain. Biol. Philos. 31, 1–24.

Bussy, T.J., Saksida, L.M., Murray, E.A., 2005. The perceptual-mnemonic/feature conjunction model of perirhinal cortex function. Q. J. Exp. Psychol. B 58, 378–396.

Cabeza, R., Locantore, J.K., Anderson, N.D., 2003. Lateralization of prefrontal activity during episodic memory retrieval: evidence for the production-monitoring hypothesis. J. Cogn. Neurosci. 15, 249–259.

Cabeza, R., Ciaramelli, E., Olson, I.R., Moscovitch, M., 2008. The parietal cortex and episodic memory: an attentional account. Nat. Rev. Neurosci. 9, 613–625.

Cabeza, R., Ciaramelli, E., Moscovitch, M., 2012. Cognitive contributions of the ventral parietal cortex: an integrative theoretical account. Trends Cogn. Sci. 16, 338–352.

Cabeza, R., Moscovitch, M., 2013. Memory systems, processing modes, and components: functional neuroimaging evidence. Perspect. Psychol. Sci. 8, 49–55.

Craver, C.F., 2016. The explanatory force of network models. Philos. Sci. 83 (5), 698–709.

Davis, S.W., Dennis, N.A., Buchler, N.G., White, L.E., Madden, D.J., Cabeza, R., 2009. Assessing the effects of age on long white matter tracts using diffusion tensor tractography. Neuroimage 46, 530–541.

Davis, S.W., Stanley, M.L., Moscovitch, M., Cabeza, R., 2017a. Resting-state networks do not determine cognitive function networks: a commentary on Campbell and Schacter. Lang. Cogn. Neurosci. 32, 669–673.

Davis, S.W., Luber, B., Murphy, D.L.K., Lisanby, S.H., Cabeza, R., 2017b. Frequency-specific neuromodulation of local and distant connectivity in aging and episodic memory function. Hum. Brain Mapp. 38, 5987–6004.

Dennis, N.A., Hayes, S.M., Prince, S.E., Madden, D.J., Huettel, S.A., Cabeza, R., 2008. Effects of aging on the neural correlates of successful item and source memory encoding. J. Exp. Psychol. Learn. Mem. Cogn. 34, 791–808.

Eichenbaum, H., Otto, T., Cohen, N.J., 1994. Two functional components of the hippocampal memory system. Behav. Brain Sci. 17, 449–472.

Eldridge, L.L., Engel, S.A., Zeineh, M.M., Bookheimer, S.Y., Knowlton, B.J., 2005. A dissociation of encoding and retrieval processes in the human hippocampus. J. Neurosci. 25, 3280–3286.

Friston, K.J., 2011. Functional and effective connectivity: A review. Brain Connect. 1, 13–36.

Friston, K.J., Buechel, C., Fink, G.R., Morris, J., Rolls, E., Dolan, R.J., 1997. Psychophysiological and modulatory interactions in neuroimaging. Neuroimage 6, 218–229.

Friston, K.J., Frith, C.D., Frackowiak, R.S., Turner, R., 1995. Characterizing dynamic brain responses with fMRI: a multivariate approach. Neuroimage (2)166–172.

Geib, B.R., Stanley, M.L., Wing, E.A., Laurienti, P.J., Cabeza, R., 2017a. Hippocampal contributions to the large-scale episodic memory network predict vivid visual memories. Cereb. Cortex 27, 680–693.

Geib, B.R., Stanley, M.L., Dennis, N.A., Woldorff, M.G., Cabeza, R., 2017b. From hippocampus to whole-brain: the role of integrative processing in episodic memory retrieval. Hum. Brain Mapp. 38, 2242–2249.

Grady, C.L., McIntosh, A.R., Craik, F.I., 2003. Age-related differences in the functional connectivity of the hippocampus during memory encoding. Hippocampus 13, 572–586.

Huneman, P., 2010. Topological explanations and robustness in biological sciences. Synthese 177, 213–245.

Iidaka, T., Sadato, N., Yamada, H., Tonekura, Y., 2000. Functional asymmetry of human prefrontal corex in verbal and non-verbal episodic memory as revealed by fMRI. Cogn. Brain Res. 9, 73–83.

Ishai, A., Ungerleider, L.G., Martin, A., Schouten, J.L., Haxby, J.V., 1999. Distributed representations of objects in the human ventral visual pathway. Proc. Natl. Acad. Sci. U. S. A. 96, 9379–9384.

King, D.R., de Chastelaine, M., Elward, R.L., Wang, T.H., Rugg, M.D., 2015. Recollection-related increases in functional connectivity predict individual differences in memory accuracy. J. Neurosci. 35, 1763–1772.

Klein, C., 2010. Philosophical issues in neuroimaging. Philos. Compass 5, 186–198.

Kornblith, S., Tsao, D.Y., 2017. How thoughts arise from sights: Inferotemporal and prefrontal contributions to vision. Curr. Opin. Neurobiol. 46, 208–218.

Leal, S.L., Noche, J.A., Murray, E.A., Yassa, M.A., 2017. Age-related individual variability in memory performance is associated with amygdala-hippocampal circuit function and emotional pattern separation. Neurobiol. Aging 49, 9–19.

Libby, L.A., Ekstrom, A.D., Ragland, J.D., Ranganath, C., 2012. Differential connectivity of perirhinal and parahippocampal cortices within human hippocampal subregions revealed by high-resolution functional imaging. J. Neurosci. 32, 6550–6560.

Majaj, N.J., Hong, H., DiCarlo, J.J., 2015. Simple learned weighted sums of inferior temporal neuronal firing rates accurately predict human core object recognition performance. J. Neurosci. 35, 13402–13418.

Medaglia, J.D., Lynall, M.E., Bassett, D.S., 2015. Cognitive network neuroscience. J. Cogn. Neurosci. 27, 1471–1491.

Meunier, D., Fonlupt, P., Saive, A.-L., Plailly, J., Ravel, N., Royet, J.-P., 2014. Modular structure of functional networks in olfactory memory. Neuroimage 95, 264–275.

Moscovitch, M., Cabeza, R., Winocur, G., Nadel, L., 2016. Episodic memory and beyond: the hippocampus and neocortex in transformation. Annu. Rev. Psychol. 67, 105–134.

Ojemann, G.A., Ojemann, J., Ramsey, N.F., 2013. Relation between functional magnetic resonance imaging (fMRI) and single neuron, local field potential (LFP) and electrocorticography (EcoG) activity in human cortex. Front. Hum. Neurosci. 7, 34.

Paolini, B.M., Laurienti, P.J., Norris, J., Rajeski, W.J., 2014. Meal replacement: calming the hot-state brain network of appetite. Front. Psych. 5, 349.

Pessoa, L., 2014. Understanding brain networks and brain organization. Phys. Life Rev. 11, 400–435.

Rathkopf, C., 2015. Network representation and complex systems. Synthese 1–24.

Rissman, J., Gazzaley, A., D'Esposito, M., 2004. Measuring functional connectivity during distinct stages of a cognitive task. Neuroimage 23, 752–756.

Rubinov, M., Sporns, O., 2010. Complex network measures of brain connectivity: uses and interpretations. Neuroimage 52, 1059–1069.

Rugg, M.D., Vilberg, K.L., 2013. Brain networks underlying episodic memory retrieval. Curr. Opin. Neurobiol. 23, 255–260.

Schedlbauer, A.M., Copara, M.S., Watrous, A.J., Ekstrom, A.D., 2014. Multiple interacting brain areas underlie successful spatiotemporal memory retrieval in humans. Sci. Rep. 4, 6431.

Shimamura, A.P., 2010. Bridging psychological and biological science: the good, bad, and ugly. Perspect. Psychol. Sci. 5, 772–775.

Simons, J.S., Peers, P.V., Mazuz, Y.S., Berryhill, M.E., Olson, I.R., 2010. Dissociation between memory accuracy and memory confidence following bilateral parietal lesions. Cereb. Cortex 20, 479–485.

Smith, S.M., 2012. The future of FMRI connectivity. Neuroimage 62, 1257–1266.

Sneve, M.H., Grydeland, H., Nyberg, L., Bowles, B., Amlien, I.K., Langnes, E., Walhovd, K.B., Fjell, A.M., 2015. Mechanisms underlying encoding of short-lived versus durable episodic memories. J. Neurosci. 35, 5202–5212.

Spaniol, J., Davidson, P.S., Kim, A.S., Han, H., Moscovitch, M., Grady, C.L., 2009. Event-related fMRI studies of episodic encoding and retrieval: meta-analyses using activation likelihood estimation. Neuropsychologia 47, 1765–1779.

Sporns, O., 2011. Networks of the Brain. MIT Press, Cambridge, MA.

Sporns, O., 2014. Contributions and challenges for network models in cognitive neuroscience. Nat. Neurosci. 17, 652–660.

Stanley, M.L., Moussa, M.N., Paolini, B.M., Lyday, R.G., Burdette, J.H., PJ, L., 2013. Defining nodes in complex brain networks. Front. Comput. Neurosci. 7, 169.

Staresina, B.P., Fell, J., Do Lam, A.T.A., Axmacher, N., Henson, R.N., 2012. Memory signals are temporally dissociated in and across human hippocampus and perirhinal cortex. Nat. Neurosci. 15, 1167–1173.

Staresina, B.P., Cooper, E., Henson, R.N., 2013a. Reversible information flow across the medial temporal lobe: the hippocampus links cortical modules during memory retrieval. J. Neurosci. 33, 14184–14192.

Staresina, B.P., Fell, J., Dunn, J.C., Axmacher, N., Henson, R.N., 2013b. Using state-trace analysis to dissociate the functions of the human hippocampus and perirhinal cortex in recognition memory. Proc. Natl. Acad. Sci. U. S. A. 110, 3119–3124.

Telesford, Q.K., Simpson, S.L., Burdette, J.H., Hayasaka, S., Laurienti, P.J., 2011. The brain as a complex system: using network science as a tool for understanding the brain. Brain Connect. 1, 295–308.

Tulving, E., 2002. Episodic memory: from mind to brain. Annu. Rev. Psychol. 53, 1–25.

Vilberg, K.L., Rugg, M.D., 2008. Memory retrieval and the parietal cortex: a review of evidence from a dual-process perspective. Neuropsychologia 46, 1787–1799.

Watrous, A.J., Tandon, N., Conner, C.R., Pieters, T., Ekstrom, A.D., 2013. Frequency-specific network connectivity increases underlie accurate spatiotemporal memory retrieval. Nat. Neurosci. 16, 349–356.

Westphal, A.J., Wang, S., Rissman, J., 2017. Episodic memory retrieval benefits from a less modular brain network organization. J. Neurosci. 37, 3523–3531.

Worsley, K.J., Friston, K.J., 1995. Analysis of fMRI time-series revisited—again. Neuroimage 2, 173–181.

FURTHER READING

Simons, J.S., Spiers, H.J., 2003. Prefrontal and medial temporal lobe interactions in long-term memory. Nat. Rev. Neurosci. 4, 637–648.

Sporns, O., 2013. Structure and function of complex brain networks. Dialogues Clin. Neurosci. 15, 247–262.

Watrous, A.J., Ekstrom, A.D., 2014. The spectro-contextual encoding and retrieval of episodic memory. Front. Hum. Neurosci. 8, 75.

Index

Note: Page numbers followed by *f* indicate figures and *t* indicate tables.

Printed in the United States
By Bookmasters